A GUIDE TO PHOSPHOLIPID CHEMISTRY

A GUIDE TO PHOSPHOLIPID CHEMISTRY

DONALD J. HANAHAN

New York   Oxford
Oxford University Press
1997

Oxford University Press

Oxford New York Athens Auckland Bangkok Bogota Bombay
Buenos Aires Calcutta Cape Town Dar es Salaam Delhi Florence
Hong Kong Istanbul Karachi Kuala Lumpur Madras Madrid Melbourne
Mexico City Nairobi Paris Singapore Taipei Tokyo Toronto

and associated companies in
Berlin Ibadan

Copyright © 1997 by Oxford University Press, Inc.

Published by Oxford University Press, Inc.
198 Madison Avenue, New York, New York 10016

Oxford is a registered trademark of Oxford University Press

Library of Congress Cataloging-in-Publication Data

Hanahan, Donald J. (Donald James), 1919–
    A guide to phospholipid chemistry / by Donald J. Hanahan.
        p. cm.
    Includes bibliographical references and index.
    ISBN 0-19-507980-9; — ISBN 0-19-507981-7 (pbk.)
    1. Phospholipids. I. Title.
QP752.P53H36 1996
591.19′247—dc20
96-7089

9 8 7 6 5 4 3 2 1

Printed in the United States of America
on acid-free paper

To Lillian,

a constant and always

supportive companion

PREFACE

Several years ago, I authored a book, together with Irving Zabin and Frank Gurd, entitled simply *Lipide Chemistry* (Wiley & Sons, New York, 1960). The concern expressed at that time was that studies on lipids centered on establishing the etiology and cure for atherosclerosis and other lipid-related diseases. It was an understated feeling at that time that many investigators used "seriously outmoded or incompletely tested methods or concepts in their enthusiastic search to a solution to these perplexing problems." In essence, there was a concern that many students (and senior investigators as well) appeared not to have any in-depth training in lipid chemistry or biochemistry. Now after some 37 years, the concern is the same but the pendulum has swung in other directions. There has been the dramatic emergence of molecular biology and its implied goal of cure of all diseases through cloning and gene manipulation, along with the revolutionary changes in our understanding of membrane biochemistry and biology. Certainly the latter is an exciting area in a personal sense because it focuses attention on the behavior of membrane lipids (primarily phosphoglycerides) and now shows that these lipids are important components of the signal transduction system initiated when an agonist interacts with a sensitive cell. Even a casual glance at most biochemically oriented journals will show an enormous number of publications in which many different methods are used to explore changes in membrane lipids subsequent to the agonist interaction. However, in reading the literature it is my personal feeling that many investigators do not really have a basic understanding of the nuances of lipid chemistry and biochemistry. In a way, that is understandable because very few major university settings (main campus, medical school, etc.) have faculty with an expertise in the area of lipid chemistry and biochemistry. There are probably less than ten major universities in this country in which graduate courses in the chemistry and biochemistry of lipids per se are offered. Essentially we are at the nadir of training of persons with a true expertise and/or knowledge in lipids and yet we have an explosive interest in these compounds in biological reactions. A perplexing situation which one hopes will change to the better in the future! Thus a large majority of students emerging in the biological sciences are never exposed to this discipline.

The primary goal of this book is to provide a gentle introduction to phospholipid chemistry, particularly as it might be applied to a study of these compounds in mammalian cells. Inasmuch as the authors' research interest over the past several years has been intimately involved with the role of phospholipids in the signal transduction process using rabbit or human platelets as a model cell, most experimental examples center on this cell. The first three chapters are devoted to an orientation or introduction to the chemical nature of lipids in general, how they are thought to be associated in the cell, and the methodology by which the cellular lipids including the phospholipids can be recovered from cells and subjected to an initial identification. Then the subsequent chapters are concerned with the chemical characteristics and characterization of the choline-containing phospholipids, including the sphingophospholipids and then the major non-choline-containing phospholipids and finally the so-called minor phospholipids. These latter compounds, though low in concentration in cells, are very high in visibility because they form the vanguard of a new category of biologically active substances. As shall be mentioned often in the main text, the finding of biologically active phospholipids (that act as agonists or lipid chemical mediators on cells) has set the stage for the study of cellular phospholipids on a new and exciting course. Finally it should be emphasized that this book is intended to provide a basis for further inquiry by an individual investigator on these complicated molecules and to show that these compounds are unique but, with care and understanding, can be studied with ease. The text follows in general the format of a graduate course given to first- and second-year graduate students in our Ph.D. degree track here. In any event it is hoped that this book will be of benefit to graduate students, postdoctoral fellows, and interested senior investigators.

The author wishes to acknowledge the generous and continued support of Dr. Merle S. Olson during the lengthy birth of this book. A debt of gratitude is owed to Esther L. Hall, whose wise, compassionate, and dedicated help brought this book to fruition.

San Antonio, Texas   D.J.H.

CONTENTS

A GUIDE TO
PHOSPHOLIPID
CHEMISTRY

ONE

INTRODUCTION TO LIPIDS

Prologue

Concomitant with the explosive development and progress made in the field of molecular biology, initiated by development and proof of the double helix structure of DNA some 40 years ago, there has been a much more subdued, but equally exciting and important, development in the area of signal transduction (Nishizuka, 1992; Berridge, 1984; Exton, 1990). What makes the latter subject so enchanting to biochemists is the finding that membrane phospholipids are intimately involved in the transduction process. Interestingly, the potential role of phospholipids in the signal transduction pathway was formulated some 40 years ago also. (Perhaps all great discoveries occur in 40 year cycles.) In any event the first hint of any possible involvement of phospholipids in cellular stimulus responses was gained from the work of Hokin and Hokin, which first appeared in 1953. In this classic paper, these investigators reported that treatment of pigeon pancreas slices with acetylcholine or carbamylcholine (cholinergic drugs) resulted not only in the secretion of amylases but also in the turnover of two specific membrane phospholipids, namely, phosphatidylinositol and phosphatidic acid. While the entire process of stimulus response in a mammalian cell is now much more complicated, nevertheless the findings by Hokin and Hokin were of major importance in the maturation of this field.

Unfortunately, the impact of the Hokins' observations was not immmediately felt. At that point in time, phospholipids were viewed mainly as semipermeable membrane structures whose main function was to regulate the ion content of the cell. In addition, another deterrent was the limited information on the chemical structure of the mammalian cell phospholipids. Hence there was a hiatus of many years in which low-profile lipid chemists and biochemists labored to solve the chemical nature of membrane lipids and to deduce their physical arrangement in the cell. Then in 1975, Michell published a key paper (Michell, 1975) in which he noted the importance of the inositol-containing phospholipids in the membrane process known as *calcium gating*. This paper initiated what can be called the "PI" era, which is still very much alive and well today.

This brief historical sketch serves as an introduction to the main goal of this book, which is to describe in some detail the chemical nature of phospholipids present in mammalian cells. It is hoped that there will be sufficient information for the reader to appreciate the uniqueness of many of these compounds, develop some rapport with their chemical structure, and become familiar with their isolation and identification.

This book will not be encyclopedic in nature and will not approach the status of a tome but rather will provide a gentle introduction to a fascinating group of compounds. Prior to a discussion of specific phospholipids, it seems worthwhile to provide some general background, definitions, and approaches used in obtaining these lipids from a cell. First, it is appropriate to consider some general facts about lipids per se.

Lipids: What Are They?

Lipids are a complex group of substances, which include the long-chain fatty acids and their derivatives, sterols and steroids, carotenoids, and other related isoprenoids. It is evident that the term *lipid* denotes a wide range of compounds that appear to have little obvious interrelation. However, although these compounds possess widely different structures, they are derived in part from similar biological precursors and exhibit similar physical and chemical characteristics. Furthermore, most lipids occur naturally in close association with protein, either in membranes as insoluble lipid–protein complexes or as soluble lipoproteins of the plasma.

Some of the lipids are of considerable importance as energy sources. Hence, oxidation of long-chain fatty acids, stored in the adipose tissue of mammals as triacylglycerols (triglycerides), may supply as much as 80% to 90% of the total energy requirements under certain circumstances. The close relationship of this metabolic path to that of carbohydrate utilization is well established. Independent of their use as fuels, the group of biologically active lipids includes the steroid hormones (estrogens, androgens, and adrenocortical hormones), other sterols (bile acids, vitamin D), and various terpenoids [β-carotene, retinal (vitamin A aldehyde)], all of which are essential components of specific metabolic processes.

In any attempt to evaluate adequately the activity or role of lipids in biological systems, it is of prime importance to acquire a firm understanding of their chemistry. This discussion will emphasize the chemical characteristics

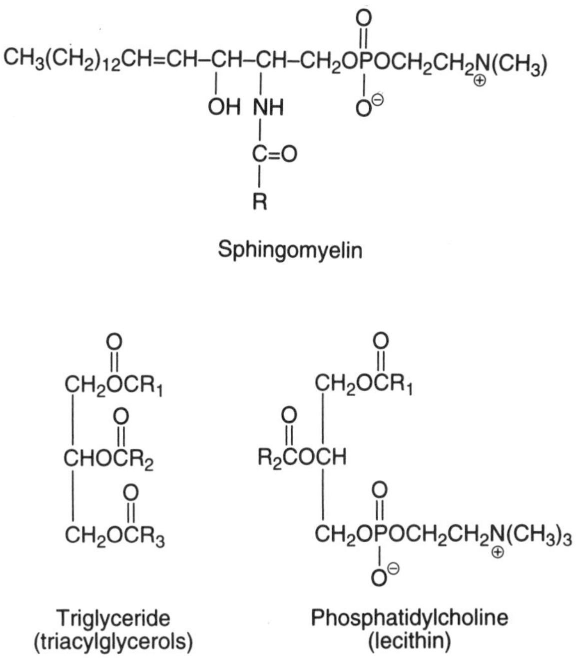

$$CH_3(CH_2)_{12}CH=CH-\underset{\underset{OH}{|}}{CH}-\underset{\underset{NH}{|}}{CH}-CH_2O\overset{\overset{O}{||}}{P}OCH_2CH_2\overset{\oplus}{N}(CH_3)$$

$$\underset{\underset{R}{|}}{\underset{\underset{C=O}{|}}{}}\qquad\underset{O^{\ominus}}{}$$

Sphingomyelin

$$\begin{array}{c}\overset{\overset{O}{||}}{CH_2OCR_1}\\ |\quad\overset{O}{||}\\ CHOCR_2\\ |\quad\overset{O}{||}\\ CH_2OCR_3\end{array}\qquad\begin{array}{c}\overset{\overset{O}{||}}{CH_2OCR_1}\\ \overset{O}{||}\quad|\\ R_2COCH\\ |\quad\overset{O}{||}\\ CH_2OPOCH_2CH_2\overset{\oplus}{N}(CH_3)_3\\ |\\ O^{\ominus}\end{array}$$

Triglyceride
(triacylglycerols)

Phosphatidylcholine
(lecithin)

FIGURE 1-1. Structural formulae of three common cellular lipids.

of naturally occurring phospholipids and the usefulness of such information in the interpretation of the biological behavior of these compounds.

In considering the distribution of particular lipids, it must be emphasized that there is wide variation in the lipid composition of various cells (see section entitled "An Excursion into the Complexities of Phospholipids Found in Certain Cells: Defining the Problem"). Of the three classes of lipids depicted in Figure 1-1, triglycerides (triacylglycerols) form the chief lipid constituent of adipose tissue in the mammal and also are found in plasma.

The closely related derivative, phosphatidylcholine, occurs in most plant and animal cells and is the most abundant lipid in mammalian cells. Another type of lipid is represented by sphingomyelin, which is found in high concentration in brain tissue, other nervous tissue, and erythrocytes.

A casual examination of the general chemical composition of membranes from mammalian cells reveals that over 90% of their (dry weight) mass is comprised of proteins and phospholipids (in the main). The weight ratio of these two major classes of compounds may vary considerably from one source to another, ranging from 3:1 in myelin to 1:3 in mitochondria. While no definitive figure can be cited for the minimal amount of phospholipid and/or protein required for a membrane structure, it is abundantly clear that any ultimate disclosure of the architecture and behavior of biological membranes must clearly include an in-depth understanding of the chemistry of these two important classes of compounds.

Thus, it is the intent to limit attention here to the phospholipids and to their interrelation with other components of membranes and with each other; examples of their participation in biologically important reactions will be explored. Prior to an in-depth treatment of the chemistry of the phospholipids, it seemed appropriate to describe some general facets of their biochemistry, especially with regard to approaches to isolation, purification, structure proof, and so on. In addition, it is appropriate to include a very brief resume of the types of fatty acids commonly found in naturally occurring lipids because it will complement later discussions on the complex phospholipids.

Fatty Acids: Very Vital Constituents

On a purely quantitative basis there can be no doubt that the fatty acids are one of the important constituents of the lipids of all living cells. The term *fatty acid* is applied to the monobasic carboxylic acids of the general formula

$$\overset{\text{O}}{\overset{\|}{\text{RC}}}\text{-OH},$$

where R may be a straight-chain saturated (alkyl) or an unsaturated (alkenyl) hydrocarbon residue; in the more prevalent naturally occurring fatty acids the R group may vary in chain length from C_{11} to C_{23} and may contain in addition hydroxy, keto, branched, or even alicyclic substitiuents. Inasmuch as it is not possible to consider here in depth all of the naturally occurring fatty acids, emphasis will be placed on the more commonly encountered types.

A more detailed description of the fatty acids may be gained from reference sources (Markeley, 1960, 1961, 1964; Kuksis, 1978). It should be emphasized, however, that fatty acids, in the free form, exist in very low levels, with the main forms being carboxylic esters or amides.

Saturated

The most representative saturated fatty acids found in animal, plant, and, to a lesser extent, bacterial lipids are palmitic acid and stearic acid (Figure 1-2).

The systematic names for these acids are given in parentheses; the systematic terms are preferable to the trivial names when it is necessary to describe the geometric isomerism of an unsaturated fatty acid or the exact location of a substituent group.

Palmitic and stearic acids are the major saturated fatty acid constituents of most animal and plant tissues. Much smaller amounts of other saturated fatty acids are present in most natural sources. Low concentrations of myristic acid (*n*-tetradecanoic acid; 14:0) and lauric acid (*n*-dodecanoic acid; 12:0) have been detected in certain tissues.

Higher fatty acids such as lignoceric acid (24:0) and behenic acid (22:0) are found in high concentrations in brain sphingolipids. Finally, evidence exists for

$$CH_3(CH_3)_{14}COOH \qquad CH_3(CH_2)_{16}COOH$$

Palmitic acid   Stearic acid

(n-Hexadecanoic acid)   (n-Octadecanoic acid)

FIGURE 1-2. Long-chain saturated fatty acids found in mammalian cells.

the occurrence of fatty acids with an odd number of carbon atoms, such as pentadecanoic acid (15:0) and heptadecanoic acid (17:0); however, these latter acids are present in very low concentrations.

A list of the more common saturated straight-chain fatty acids, which gives their usual sources, is presented in Table 1-1.

It is important to emphasize that lower homologs of these and other fatty acids are also of biological significance. Specifically, the shorter-chain saturated fatty acids [e.g., butyric acid (4:0) and caproic acid (6:0)], are important constituents of milk lipids, and octanoic acid (8:0) and decanoic acid (10:0) are present in high concentrations in palm oil. The short-chain fatty acids are rarely found in mammalian organs or tissues—with the exception of milk, where they are in relatively high concentration.

Unsaturated

These long-chain fatty acids have formulae similar to those described above but contain one or more double bonds. This group of fatty acids may be subdivided into monoenoic and polyenoic types. A general summary of the chemical nature of the unsaturated fatty acids follows.

MONOENOIC. (Monoethenoic, Monounsaturated, Ethylenic, Alkenoic). Although acids containing olefinic bonds (double bonds) or acetylenic bonds (triple bonds) have been isolated from natural sources, most attention will be directed toward the former types.

The commonly occurring unsaturated fatty acids contain a single double bond. There are a large number of possible monoenoic acids; not only are

TABLE 1-1. Some of the More Common Saturated Straight-Chain Fatty Acids

Trivial Name	Systematic Name	Chain Length in Carbon Atoms	Typical Source
Lauric acid	n-Dodecanoic acid	12	Palmkernel oil, nutmeg
Myristic acid	n-Tetradecanoic acid	14	Palmkernel oil, nutmeg
Palmitic acid	n-Hexadecanoic acid	16	Olive oil, animal fat
Stearic acid	n-Octadecanoic acid	18	Cocoa butter, animal, fat
Behenic acid	n-Docosanoic acid	22	Brain, radish oil
Lignoceric acid	n-Tetracosanoic acid	24	Brain, carnauba wax

$$CH_3(CH_2)_5CH=CH(CH_2)_7COOH$$

Palmitoleic acid

(cis-9-Hexadecenoic acid)

$$CH_3(CH_2)_7CH=CH(CH_2)_7COOH$$

Oleic acid

(cis-9-Octadecenoic acid)

FIGURE 1-3. Representative long-chain monounsaturated fatty acids.

there many possible positions for the olefinic bond, but geometric isomers can also occur. These considerations plus the fact that fatty acids of different chain lengths exist make the problem appear immense. However, relatively few of the theoretically possible monoenoic acids occur in animal and plant tissues. Thus, the major and most representative monoenoic acids occur in animal and plant tissues. The major and most representative monoenoic acids present in animal and plant tissue are oleic acid and palmitoleic acid (Figure 1-3).

The structures of these fatty acids are characterized by (a) the presence of an olefinic bond between carbon atoms 9 and 10 (counting from the carboxyl end) and (b) the occurrence of a *cis* configuration rather than a *trans* configuration at the double bond. Fatty acids are known in which a single double bond occurs at other positions and in which a *trans* configuration is present, but these are relatively rare.

A list of the common monoenoic fatty acids is given in Table 1-2.

POLYENOIC.   [Polyethenoic, Polyunsaturated, Alkapolyenoic (Alkadienoic, Alkatrienoic, etc.)]. In contrast to the monounsaturated fatty acids, these unsaturated fatty acids are found in much smaller amounts in naturally occurring lipids. The structural formulas of the three most commonly encountered polyunsaturated fatty acids are in Figure 1-4. Linoleic acid and arachidonic

TABLE 1-2. Common Monoenoic Fatty Acids

Trivial Name	Systematic Name	Chain Length in Carbon Atoms	Typical Source
Palmitoleic acid	*cis*-9-Hexadecenoic acid	16	Marine algae, pine oil
Oleic acid	*cis*-9-Octadecenoic acid	18	Animal tissue, olive oil
Gadoleic acid	*cis*-9-Eicosenoic acid	20	Fish oils (cod, sardine)
Erucic acid	*cis*-13-Docosenoic acid	22	Rapeseed oil
Nervonic acid	*cis*-15-Tetracosenoic acid	24	Elasmobranch fish, brain

$$CH_3(CH_2)_4CH=CHCH_2CH=CH(CH_2)_7COOH$$

Linoleic acid

(cis-9, cis-12-octadecadienic acid)

$$CH_3CH_2CH=CHCH_2CH=CHCH_2CH=CH(CH_2)_7COOH$$

Linolenic acid

(cis-9, cis-12, cis-15-octadecatrienoic acid)

$$CH_3(CH_2)_4CH=CHCH_2CH=CHCH_2CH=CHCH_2CH=CH(CH_2)_3COOH$$

Arachidonic acid

(cis-5, cis-8, cis-11, cis-14 eicosatetraenoic acid)

FIGURE 1-4. Primary long-chain polyunsaturated fatty acids.

acid form the bulk of the polyunsaturated fatty acids in most animal tissues. As evident from the structures given in Figure 1-4, these acids possess more than one center of geometric isomerism (as the double bonds) and hence many isomeric forms are possible. An examination of the polyunsaturated fatty acids found in nature shows that there is a marked prevalence of acids possessing the following characteristics: (a) nonconjugated double bonds (frequently termed a divinylmethane rhythm, as in linoleic acid); (b) double bonds at specific positions, as in linoleic and arachidonic acids; and (c) *cis* configuration at all the double bonds.

A series of the more frequently encountered and relatively important polyunsaturated fatty acids is listed in Table 1-3.

TABLE 1-3. Series of the More Frequently Encountered and Relatively Important Polyunsaturated Fatty Acids

Trivial Name	Systematic Name	Chain Length in Carbon Atoms	Typical Source
Linoleic acid	*cis*-9, *cis*-12-Octadecadienoic acid	18	Corn oil, animal tissue, bacteria
Linolenic acid	*cis*-9, *cis*-12, *cis*-15-Octadecatrienoic acid	18	Animal tissues
—	5,8,11-Eicosatrienoic acid	20	
—	8,11,14-Docosatrienoic acid	20	Brain
—	7,10,13-Docosatrienoic acid	22	Phospholipids
—	8,11,14-Docosatrienoic acid	22	
Arachidonic acid	5,8,11,14-Eicosatetraenoic acid	20	Liver, brain
—	4,7,10,13-Docosatetraenoic acid	22	Brain
—	4,7,10,13,16,19-Docosahexaenoic acid	22	Brain

It is noteworthy that virtually all of the naturally occurring polyunsaturated fatty acids contain 18–22 carbon atoms. Linoleic acid is the predominant polyunsaturated fatty acid. It is of considerable interest that most animals cannot synthesize linoleic acid and must take it in the diet. If insufficient amounts of this acid are present in the diet of animals, severe symptoms, such as skin lesions, kidney damage, cataracts, increased permeability to water, and so on, can occur. Thus, the term *essential fatty acids* has been applied to these compounds. It is not certain, however, that dietary unsaturated fatty acids are needed by the human adult, but there is evidence of such a requirement by the human infant.

Summary

This brief excursion into the chemical nature of naturally occurring fatty acids should attest to their diverse structural characteristics. In general, the aforementioned fatty acids occur only to a limited extent in the free form but occur more frequentyl in the ester or amide form. The possible number of permutations in fatty-acid-containing species are impressive and provides an interesting and challenging problem. Certain fatty acids tend to be identified rather characteristically with specific lipids, namely, phosphoglycerides and sphingolipids. Furthermore, a close examination of the positioning of fatty acids on certain phosphoglycerides reveals a high degree of specificity. Thus, most mammalian tissues—phosphatidyl choline, for example—contains primarily saturated fatty acids on the C-1 position and primarily unsaturated fatty acids on the C-2 position. Other phospholipids tend to have similar structural patterns with regard to positioning of fatty acids or in the specific types of fatty acids present.

A Comment on Nomenclature

Phosphoglyceride refers to a glycerophosphoric acid derivative that contains a minimum of one *O*-acyl, *O*-alkyl, or *O*-alk-1-enyl group covalently linked to the glycerol backbone. An example would be 1,2-diacyl-*sn*-glycero-3-phosphoric acid, which is often referred to in the literature as *phosphatidic acid*. The latter (with one less hydrogen on the phosphate group) is frequently called a *phosphatidyl group,* which is used to define, for example, a derivative with a choline group esterified to the phosphoric acid moiety on a phosphatidic acid. Hence the derivation of the term *phosphatidylcholine.* The use of this type of terminology is widespread and is very convenient. Of course, if the configuration is known, one could use the term 3-*sn*-phosphatidylcholine or 1,2-diacyl-*sn*-glycero-3-phosphocholine.

Phospholipid is a generic term that refers to lipids containing a phosphoric acid residue. Thus, both phosphatidylcholine and sphingomyelin would fit in this category.

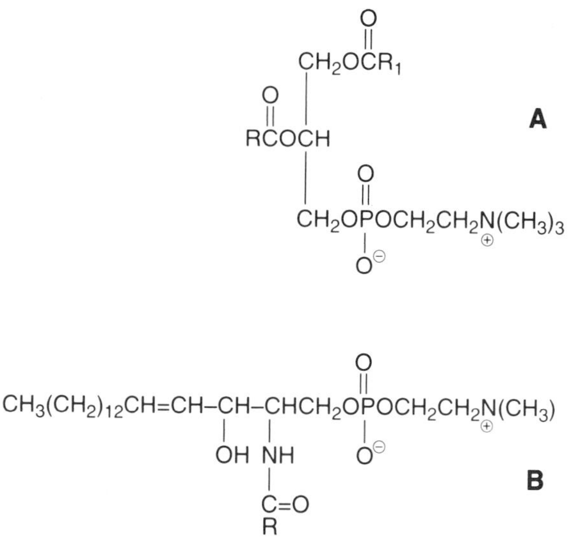

$$\begin{array}{c} O \\ \parallel \\ CH_2OCR_1 \\ O \quad | \\ \parallel \; | \\ RCOCH \\ | \quad\quad O \\ \quad\quad \parallel \\ CH_2OPOCH_2CH_2N(CH_3)_3 \\ | \\ O^{\ominus} \end{array}$$

A

$$CH_3(CH_2)_{12}CH=CH-CH-CHCH_2OPOCH_2CH_2N(CH_3)$$
$$\begin{array}{ccc} | & | & | \\ OH & NH & O^{\ominus} \\ & | & \\ & C=O & \\ & R & \end{array}$$

B

FIGURE 1-5. Two phospholipids commonly encountered in the cellular membranes.

Some Basic Facets of Phospholipid Structure

The More Prevalent Forms

It it appropriate at this point to provide a general description of the types of phospholipids one encounters in mammalian cells and those which we will be considering in this book. Essentially there are two major classes of phospholipids, one of which contains a glycerol backbone and the other of which contains a sphingosine backbone. Their general structural formulae are presented in Figure 1-5.

Structural formula A is typical of a group of compounds commonly called phosphoglycerides. In this particular depiction the symbols R_1 and R_2 are long-chain carboxylic acids linked in an ester linkage to the primary and secondary alcohol residues of glycerol. The symbol X is a mixed venue with a wide variety of bases—for example, choline ($HOCH_2CH_2 N (CH_3)_3$) or

OH

ethanolamine ($HOCH_2CH_2NH_2$)—joined in ester linkage to the phosphoric acid residue.

As we shall see later, nature, as expected, is not so simple and straightforward, since variation of the structure given in A can occur. Thus a vinyl ether (O—CH=CH—R) residue or a saturated ether moiety (O—CH_2CH_2R) can replace the carboxylic acid ester on carbon 1 of glycerol. In addition, there can be at least eight different substituents associated with the phosphoric acid residue. Thus the number of permutations in structure can be enormous, perhaps as great as 1000, given variability in the substitutent at the C-1 and C-2 positions.

However, one should not dispair at this complexity since there is some order in all this chaos. First only one type of backbone residue, glycerol, is found in the most prevalent type of phospholipids, the phosphoglycerides. Even the sphingophospholipids (formula B) in mammals have only one other type (minor) of backbone group, a saturated form called *dihydrosphingosine*. Second, high stereospecificity is exhibited by the phosphoglycerides, with only one form (the *sn*-3 configuration) being found (stereospecificity will be discussed again later in this chapter). Positional specificity also predominates in the phosphoglycerides, with the C-1 position containing primarily (over 95%) saturated residues and the C-2 position having over 95% unsaturated residues. Finally, the phosphoric acid residues are found mainly (over 98%) at the C-3 alcohol position. Thus, it is rather comforting to see that there are some common denominators among these complex lipids.

Structure B shows the chemical formula of a sphingosine-based phospholipid, sphingomyelin. This compound is the most predominant form found in mammalian tissues and possesses some interesting chemical features. The backbone, in this case a long-chain amino alcohol (sphingosine), is normally found with the amino function amidated with a long-chain fatty acyl chain and with a phosphoric acid group on the alcohol group. In addition, the latter function can be esterified to a base such as choline (a prevalent form). Only a minor amount of a saturated form, dihydrosphingosine, has been detected. Some evidence for a longer-chain hydrocarbon has been presented, but again these forms are normally very minor constituents of mammalian cells. Interestingly, while there are many different chain lengths (and degrees of unsaturation) noted with the fatty acids associated with the phosphoglycerides, there are relatively few different types of fatty acids found in the sphingomyelins; these are mainly saturated with restricted chain lengths (e.g., 18:0 and 22:0). Further discussion of these compounds and more complicated 18:0 and 22:0 forms, such as gangleoside, will occur in Chapter 4.

Stereochemistry of the Phosphoglycerides

All configurational assignments rest upon the classic work of Baer and his associates. They used the D/L system, which placed monoacylglycerol in the same category as the glyceraldehyde into which it could be transformed by oxidation, without alterations or removal of any substituent (Baer and Buchnea, 1959). Steric representation of a triglyceride is thus possible as shown in Figure 1-6. Equivalent but easier to show is the structure depicted in Figure 1-7.

While an optical classification of triacylglycerols could be made, confusion arose over the two primary alcohols.

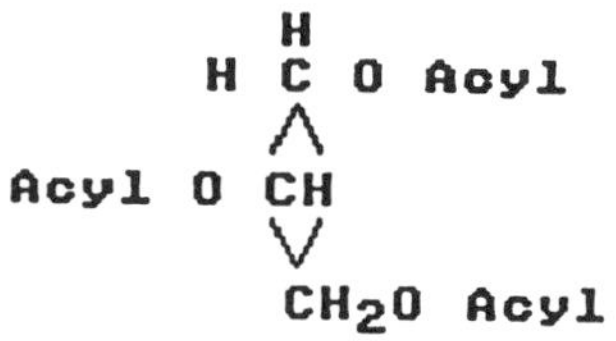

```
            H
       H    C   O  Acyl
            /\
Acyl  O  CH
            \/
          CH2O  Acyl
```

FIGURE 1-6. Stereochemical depiction of a triglyceride by a Fischer formulation.

This difficulty led to the R/S system devised by Cahn, Ingold, and Prelog (1966). This identifies substituents on the chiral C of glycerol by the clockwise (R) or counterclockwise (S) order of their priorities. In this system, the priority of substituents on the C-2 atom of a triacylglycerol is such that the oxygen atom has the highest rank while the hydrogen atom has the lowest rank. The decision on the second and third rank rests with the chain length and substitutions of fatty acids esterified to the primary position, with the long-chain fatty acids being ranked over the unsaturated and short-chain fatty acids.

Thus, description of a simple change of fatty acid at the primary position required frequent changes of configuration prefixes. Also, the R/S system, like the older D/L one, did not account for the stereospecificity of the acylglycerol derivatives toward lipases (phospholipase A_2 in particular). Finally, nonrandom distribution of fatty acids in natural or synthetic enantiomeric acylglycerols could not be systematically correlated by reference to either the R/S or D/L configuration.

Stereospecific Numbering (sn System)

In 1960, Hirschmann recognized that the two primary carbinol groups of the glycerol molecule are not interchangeable in their interactions with asymmetrical structures, including nearly all biochemical processes (Ogston, 1948). The carbinol group to receive the lowest number is a general one and is based

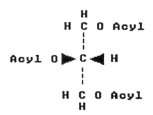

FIGURE 1-7. An alternate manner for expression of steric configuration of a triglyceride.

on the priorities of the R/S system of Cahn, Ingold and Prelog (1966). Thus, if the secondary alcohol is shown to the left of the C-2 in the Fischer projection the carbon atom above, it is called C-1 and the one below is called C-3 (IUPAC-IUB Commission, 1967). This system is widely applied to phosphoglycerides and will be used throughout this book.

Sphingolipid Stereochemistry

The "backbone" structure of the sphingolipids found in mammalian cells is the long-chain base, sphingosine. The latter is an 18-carbon moiety comprised of an aminoethanol residue at C-1 (OH) and C-2 (NH$_2$) with an allylic hydroxyl at C-3. Another name for sphingosine is D-erythro-2-amino-octadec-4-ene-1,3-diol. The latter compound is often referred to as sphingenine by many investigators.

The question of the configuration of carbons 1 and 2, which obviously contain asymmetric centers, has been addressed in several different ways. The configuration of the C-2 amino residue was proven by conversion to derivatives, one of which was N-benzoyl-α-aminostearic acid. Primarily using optical activity as a guideline, the properties of this acid could be compared to those of N-acyl-amino acids of known structure and configuration. As a result, the D configuration was assigned to carbon atom 2. Other, more elegant experimental approaches provided support for the D configuration assigned to the carbon-2 amino group. Carbon 3 is established to have an *erythro* configuration as it relates to carbon atom 2. One approach was to subject the double bond of sphingosine to ozonolysis and recover the four-carbon fragment. The latter was shown to have an *erythro* configuration through a series of experiments in which this four-carbon fragment was compared to synthetic compounds with known structure and optical configuration. It was important to eliminate the possibility of conversion to a *threo* configuration under experimental conditions. This was done in these studies, and an erythro configuration was confirmed.

On the basis of the comparison of the infrared spectrum of synthetic compounds with sphingosine from naturally occurring sources, a *trans* configuration was assigned to the C-4 double-bond system.

Consequently, as a result of this information, it is possible to depict the structure of sphingosine from natural sources, by a Fischer projection formula (Figure 1-8).

Positional Asymmetry in Phosphoglycerides

In addition to their stereochemical forms, the phosphoglycerides also exhibit another type of asymmetry in many cases, that relating to the specificity of positioning of fatty acids on the glycerol backbone. At first glance, it would seem that there would be considerable mixing of species of chain lengths and unsaturation at the C-1 and C-2 position of a molecule such as phospha-

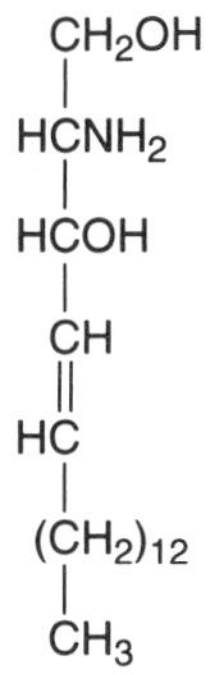

$$CH_2OH$$
$$|$$
$$HCNH_2$$
$$|$$
$$HCOH$$
$$|$$
$$CH$$
$$||$$
$$HC$$
$$|$$
$$(CH_2)_{12}$$
$$|$$
$$CH_3$$

Sphingosine

FIGURE 1-8. The predominant structural form of naturally occurring sphingosine.

tidylcholine. However, this is not the usual case since there is a high degree of positional specificity with the saturated fatty acids occupying the C-1 ester position and the unsaturated long-chain fatty acids at the C-2 position. The structural formula in Figure 1-9 will show one example.

The proof of such a distribution has derived mainly from the use of phospholipase A2, which acts stereospecifically at the C-2 ester bond of the *sn*-3 form of the phosphoglyceride. The synthesis of stereospecific forms of the phosphoglycerides has aided in establishing the specificity of the phospholipase A2. This point will be pursued in Chapter 4.

Examples of Phospholipids Found in Mammalian Cells

For general background and orientation purposes, it is worthwhile to present a brief summary of the different types of compounds.

The term, phosphatidyl, is reserved for describing phosphoglycerides which bear two fatty acyl esters in a glycerophosphoric acid backbone. How

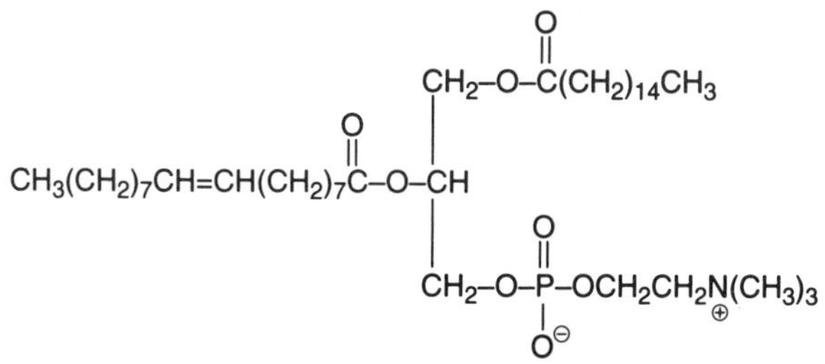

$$CH_3(CH_2)_7CH=CH(CH_2)_7\overset{\overset{\displaystyle O}{||}}{C}-O-CH$$

$$CH_2-O-\overset{\overset{\displaystyle O}{||}}{C}(CH_2)_{14}CH_3$$

$$CH_2-O-\overset{\overset{\displaystyle O}{||}}{P}-OCH_2CH_2\overset{+}{N}(CH_3)_3$$
$$\underset{O^{\ominus}}{|}$$

FIGURE 1-9. The asymmetric positioning of saturated and unsaturated fatty acids in phosphatidylcholine.

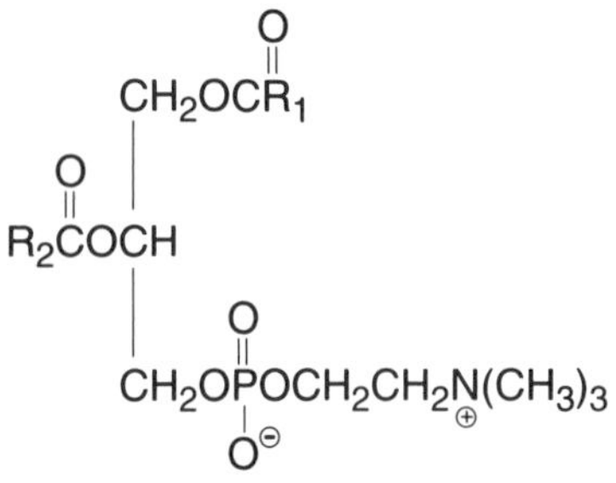

$$
\begin{array}{c}
\text{O} \\
\| \\
\text{CH}_2\text{OCR}_1
\end{array}
$$

$$
\begin{array}{c}
\text{O} \\
\| \\
\text{R}_2\text{COCH}
\end{array}
$$

$$
\begin{array}{c}
\text{O} \\
\| \\
\text{CH}_2\text{OPOCH}_2\text{CH}_2\overset{\oplus}{\text{N}}(\text{CH}_3)_3 \\
| \\
\text{O}^{\ominus}
\end{array}
$$

Phosphatidylcholine

FIGURE 1-10. The ubiquitous phosphatidylcholine molecule.

ever, the "phosphatidylcholine" fraction isolated from many mammalian cells contains two additional analogous structural components (Figure 1-10). The first is one containing a vinyl ether group (alkenyl) at the C-1 position plus a fatty acid ester substituent at the C-2 position. The second is one containing a saturated ether group (alkyl ether) at the C-1 position and a fatty acyl substituent at the C-2 position. The latter are not by definition phosphatidylcholines and, hence, must be named by their specific chemical structure.

Mixtures of the diacyl, alkenylacyl, and alkylacyl choline phosphoglycerides are found in cells such as macrophages and neutrophils, whereas in liver cells, only the diacyl form is present. To date, there have been no reports of the ether substituents being located in the *sn*-2 position, and there has been only an isolated report on the occurrence of a disaturated ether derivative in mammalian cells.

The nomenclature rules just cited for phosphatidylcholine and its analogs also hold for ethanolamine phosphoglycerides. Mixtures of the diacyl, alkenyl acyl, and alkyl ether forms are found, for example, in macrophages and neutrophils but in much different ratios than observed for the choline-containing types (Figure 1-11). Again, only the (diacyl) phosphatidylethanolamine is found in liver cells. These structures will be discussed in more detail in Chapter 5.

Phosphatidylinositol is the predominant form of inositol phospholipids found in mammalian cells, representing over 95% of the total phosphoinositides. Interestingly, no vinyl ether or saturated ether derivatives have been detected in mammalian cells. The compound shown in Figure 1-12 is found in most mammalian cells along with two very important derivatives. The latter are 1,2-diacyl-*sn*-glycero-phospho-*myo*-inositol-4 phosphoric acid [phosphatidylinositol-4-phosphoric acid (PIP)] and 1,2-diacyl-*sn*-glycero-3-phospho-*myo*-inositol-4,5-bisphosphoric acid or phosphatidylinositol-4,5-bisphosphoric acid (PIP2). This summary is of necessity limited in scope but does emphasize those compounds of interest to many different groups of investigators and will illustrate the considerable variation in structure of these naturally occurring compounds. However, this knowledge does allow us to

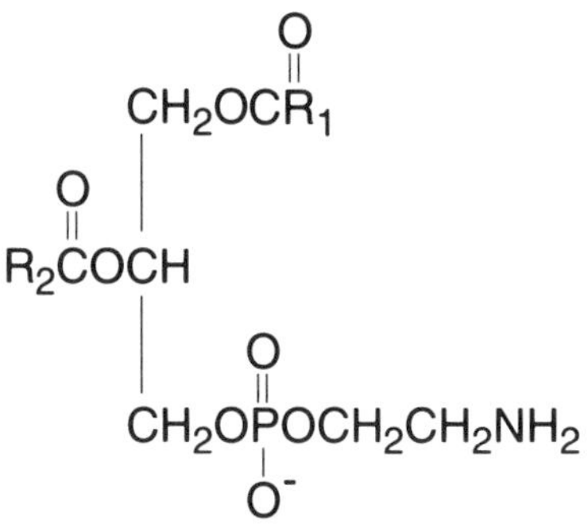

$$
\begin{array}{c}
\text{O} \\
\| \\
\text{CH}_2\text{OCR}_1 \\
\end{array}
$$

FIGURE 1-11. A major non-choline containing phosphoglyceride, phosphatidyl-ethanolamine, found in mammalian cells.

explore some constant and some not so constant features of their presence in cells.

An Excursion into the Complexities of Phospholipids Found in Certain Cells: Defining the Problem

The question raised as to the role (or behavior) of phospholipids in a biological setting centers on whether there is general uniformity of species or whether there is a random setting. There is both good news and bad news. There are certain examples of uniformity of phospholipids in various cells and there are easily an equal number of examples of nonuniformity. A brief discussion of this subject follows.

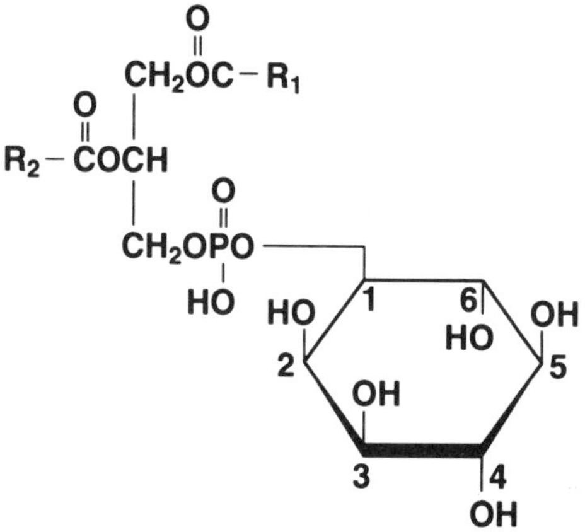

FIGURE 1-12. Phosphatidylinositol, an important phosphoglyceride precursor for membrane-active inositol phospholipids.

TABLE 1-4. Distribution of Lipids in Various Mammalian Cells (Percent of Total Lipids)

	Human Erythrocyte	Bovine Erythrocyte	Human Platelet
Cholesterol	25	30	22
Phosphatidyl ethanolamine	20	20	21.0
Phosphatidyl serine	8	4	7.8
Phosphatidyl inositol	6	4	4.8
Phosphatidyl choline	19	0	30.6
Sphingomyelin	20	42	13
Ganglioside			
Glycolipids	<1	<1	

As one further explores the chemical and compositional nature of the phospholipids in cells, some very interesting comparisons can be made, admittedly using a broad brush. The distribution of lipids (including cholesterol), human erythrocytes (red blood cells), bovine erythrocytes, and human platelets serve as good examples. This information is provided in Table 1-4.

The information in Table 1-4 illustrates significant differences in two lipids in the two erythrocyte sources. Certainly the lack of any significant amounts of phosphatidylcholine in the bovine erythrocyte is most dramatic. It is interesting to note, however, that the level of the other choline-containing phospholipid, sphingomyelin, increases in an apparent compensatory fashion. A further difference is divulged if one examines closely the composition of the phosphatidylethanolamine fraction. Again there are important differences particularly within the phosphatidylethanolamine fraction isolated from bovine erythrocytes. In this cell nearly 75% is represented by a saturated ether derivative, 1-O-alkyl-2-acyl-sn-glycero-3-phosphoethanolamine, with the remainder being the diacyl derivative. There is evidence for the presence of an N-methylated derivative, but this comprises only a very small percentage of the total ethanolamine phosphoglyceride fraction. On the other hand, in the phosphatidylethanolamine fraction of human erythrocyes there is no detectable saturated ether derivative. The major component is the diacyl phosphatidylethanolamine, with a low percentage of a vinyl ether analog, 1-O-alkenyl-2-acyl-sn-glycero-3-phosphoethanolamine. No evidence for the presence of the alkyl ether analog has been forthcoming.

There are also differences in the composition of the phospholipids in human platelets. These cells contain a little over twice as much phosphatidylcholine as sphingomyelin. Another dramatic difference is noted in the ether phospholipid content in the human platelets, where there is approximately 37% vinyl ether-containing components in the ethanolamine fraction whereas there only are 10% or less in the other phosphoglyceride species. The vinyl ether linkage has not been detected in the sphingolipids of these cells.

This brief summary on the phospholipid composition in cells will attest to a

TABLE 1-5. Molar ratios of Choline to
Non-Choline-Containing Phospholipids in Cells

Human Erythrocyte	Bovine Erythrocyte	Human Platelet
1.13	1.20	1.27

considerable diversity of lipid species. While a point will be made below that there is some constancy in the ratios of choline to noncholine species, the data presented in Table 1-2, which can be repeated in many other cell types, pose the question as to the reason for the diversity of lipid species. Again there is no clearly defined answer, especially in a cell such as the erythrocyte, whose most important function is the transport of oxygen.

While the point was made as to the variability in phospholipid composition in cells, a facet of the data presented in Table 1-4 has titillated the phagocytes of this author for some years. Basically it centers on the quite constant ratio of choline to noncholine phospholipids in cells. This is clearly evident in the data presented in Table 1-5.

Similar molar ratios have been reported in many different mammalian cells. A ready explanation for this type of distribution has not been provided to date, but it is important to note that this constancy in the ratio of choline to noncholine phosphoglycerides holds even in diseased states where the composition within a specific group may change. A particularly striking example is that shown in the erythrocytes of patients with the syndrome termed *acanthocytosis* (actually more correctly classified as *abetalipoproteinemia*). These cells exhibit a thorny (starlike) structure and have an altered lipid profile specifically in the choline-containing phospholipids (Ways, Reed, and Hanahan, 1963). In the abnormal cell, the percentage of phosphatidylcholine is typically close to 18% (of total lipid) and the sphingomyelin averages near 32%. On the other hand, in red cells from family members without this condition, the phosphatidylcholine level was near 30%, with the sphingomyelin close to 23%. Of particular importance is that even a casual examination of these data will show that the total choline-containing phospholipids were nearly identical. There were no overt differences in the composition of the non-choline-containing phosphoglycerides. While no evidence is at hand to explain the cell's compensatory mechanism, it is obvious that the cell has the ability to try to maintain a "normal" composition.

A final example of the uniqueness of phospholipid structure and composition in cells relates to the fatty acyl (or the hydrocarbon moiety) groups of particular lipid species in cells. Again this distribution, to be illustrated below, is very constant in normal cells and yet raises the question as to why specific phospholipids have a penchant for certain fatty acyl (hydrocarbon) groups which is not exhibited by other phospholipid species in the cell. This question is clearly posed in the data shown in Table 1-6. This is an abbreviated examination of only a few cells, but a similar profile occurs in many other cells as well.

This brief summary of major fatty acyl residues in the phospholipids in

TABLE 1-6. An Abbreviated Inspection of the Fatty Acids Present in Specific Phospholipids of Human Platelets (Data Given as Percentage of Total Fatty Acids)

	Phosphatidyl-choline[a]	Phosphatidyl-ethanolamine[a]	Phosphatidyl-inositol	Sphingomyelin
16:0	22	9	2	23
18:0	16	17	42	
18:2	12	2	2	[b]
20:4	14	37	33	
22:0	ND[c]	ND	ND	30
24:0	ND	ND	ND	12
24:1	ND	ND	ND	14

[a]These values include the contribution of the hydrocarbon chain associated with the alkenyl ether and alkyl ether residue found in this fraction.
[b]Less than 15% of total fatty acyl residues of this chain length.
[c]ND, non detectable.

human platelets is typical of many other cells. This type of information is especially important to investigators involved in the study of signal transduction in cells. The turnover of phosphatidylcholine, phosphatidylethanolamine, and the inositol-containing phosphoglyceride fraction plays an important role in stimulus-induced changes in cells. Also, evidence is accumulating that sphingomyelin metabolism may occur in the signal tranduction process. In any event the data in Table 1-4 show very clearly that there is specificity in the pattern of distribution of long-chain fatty acyl groups in specific phospholipids. Neither the importance of such a distribution to the cell nor the mechanism by which this specificity is achieved in a cell is understood at this point in time.

As underscored above, there is one other area of specificity of chemical structural patterns in cellular membrane phospholipids that is perplexing yet fascinating and also a consistent characteristic of particular phospholipids. This concerns the distribution of alkylacyl and alkenylacyl types of phosphoglycerides found in many cells. Table 1-7 provides data on such a distribution in human eosinophils.

Of further interest, there are no detectable alkenylacyl or alkylacyl phosphoglycerides found in the serine- or inositol-containing phosphoglycerides or in the sphingomyelin fraction.

TABLE 1-7. Comparison of the Composition of Specific Phosphoglycerides in Human Eosinophils (Data Given as Percent Total in Fraction)[a]

Group	Choline Fraction	Ethanolamine Fraction
Alkenylacyl	4	73
Alkylacyl	76	6
Diacyl	20	21

[a]These data are taken from Ojima-Uchiyama et al. (1988).

Role of Phospholipids in Cellular Processes: A Diverse Spectrum of Activities

Once the question of the presence of a membranous structure per se in mammalian cells was answered some 40 years ago through the advent of electron microscopy, many different queries were posed as to (a) the location of cellular phospholipids whether in the membranes or in the cytoplasmic fluid, or both, (b) their metabolic turnover pattern, and (c) their role in cellular processes. For a number of years it was uncertain whether phospholipids were all located in the membrane. This question was answered in the 1950s with several reports showing that all of the cell's lipid was associated with membrane structures; however, it was to be a number of years later until the concept of the dynamic participation of phospholipids in biological systems was slowly developed. As was noted earlier, a significant clue to the behavior of phospholipids in cell stimulus reactions was uncovered by Hokin and Hokin in 1954. Subsequently over the years, research findings, first in the chemistry of naturally occurring phospholipids and then in their participation in specific biological reactions, led to a "mini-revolution" in our understanding the importance of phospholipids in cell functions. No longer are these compounds viewed as passive components, mainly functioning as semipermeable barriers for cation/anion transfer. A few representative examples follow of phospholipid involvement in certain cellular events.

Cofactor Behavior

Two examples will be given in which phospholipids can be considered as cofactors which do not undergo any chemical alteration. Interestingly in both of these examples, a charged phospholipid or phospholipid mixture is required and the effect can be quite dramatic.

Human Erythrocyte (Ca^{2+}/Mg^{2+})–ATPase

This enzyme is considered to be important to maintenance of the proper level of calcium ion in the erythrocyte. As such it is also considered to be typical of a membrane-bound enzyme. Upon purification of this enzyme from human erythrocytes, it was found that the purified enzyme exhibited no activity until a mixture of a nonionic detergent and a charged phospholipid (e.g., phosphatidylserine or phosphatidylinositol), were included in the assay system (Nelson and Hanahan, 1985). Ten- to twelvefold increases in activity could be achieved. If a neutral phospholipid such as phosphatidylcholine were substituted for the phosphatidylserine (or phosphatidylinositol), essentially there was little or no enzymatic activity.

The Prothrombinase Complex

In blood coagulation in the mammal, a key final reaction is

Prothrombin $\rightarrow$ Thrombin

which involves the proteolytic cleavage of the prothrombin molecule (molecular weight ~72,000) to yield thrombin (molecular weight ~38,000), which is an active protease. The latter enzyme then attacks fibrinogen; this action essentially leads to fibrin, a major component in clot formation. It has been shown that the proteolysis of prothrombin can be catalyzed *in vitro* by a combination of factor Xa (a serine protease formed from normally inactive plasma factor X), factor V (high-molecular-weight plasma protein), phospholipids with a net negative charge, and calcium ions. Factor V is thought to act as a high-molecular-weight cofactor which avidly binds the phospholipids. A general formula for the interaction can be described as follows:

Factor V: Phospholipid $\cdot$ Ca^{2+} $\cdot$ Factor Xa

This complex can be passed through a Sephadex G-200 column in the presence of calcium ions and elutes in the void volume or close to it (depending the type of Sephadex) and exhibits potent proteolytic activity toward prothrombin. This high-molecular-weight complex is called *prothrombinase*. If this enzyme preparation is passed through a Sephadex column in the absence of calcium ions, the complex dissociates and a factor V $\cdot$ phospholipid fraction elutes separate from the factor Xa. The latter has very low proteolytic activity as isolated. However, upon recombination of these fractions in the presence of calcium ions, the high-molecular-weight prothrombinase can be recovered in highly active form. Again the phospholipid in this instance is considered to be a cofactor and is not chemically altered in any of the reactions, and its influence on the reaction is illustrated in Table 1-8.

Phospholipid Metabolic Products with Biological Activity

As has been noted before, the paper by Hokin and Hokin (1953) on stimulus influence on pancreas slices showed for the first time that activation of a tissue

TABLE 1-8. Relative Rates
of Prothrombin Activation

X_a	1
$X_a \cdot$ Ca^{2+}	3
$X_a \cdot$ Ca$^{2+} \cdot$ PL^a	27
$X_a \cdot$ Ca$^{2+} \cdot$ V	425
$X_a \cdot$ Ca$^{2+} \cdot$ V $\cdot$ PL^a	>300,000

^{a}A mixture of phosphatidylcholine and
phosphatidylserine.

led to concomitant turnover of certain (but not all) membrane phospholipids—in particular, phosphatidylinositol and phosphatidic acid. In the period since that discovery, other inositol derivatives of interest have been isolated and their structure determined. The most provocative one was phosphatidylinositol-4,5-bisphosphate (Ptdins-4,5-P_2).

Also as has been discussed before, Michell (1975) reported that the phosphoinositides (this term includes all the different phosphorylated forms) were intimately involved in the calcium gating process in cells. Current dogma states that under stimulus conditions, there is a rapid (within 5–30 sec) metabolic change in the Ptdins-4,5-P_2 through the action of a phospholipase C. Two products are formed as follows:

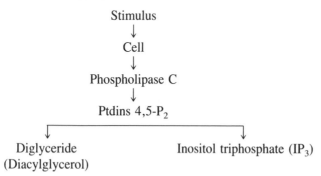

$$Stimulus$$
$$\downarrow$$
$$Cell$$
$$\downarrow$$
$$Phospholipase\ C$$
$$\downarrow$$
$$Ptdins\ 4,5\text{-}P_2$$

Diglyceride   Inositol triphosphate (IP_3)
(Diacylglycerol)

IP_3 is classified now as a second messenger with considerable influence on calcium movement from intracellular stores in a cell. Furthermore, diglycerides have a considerable and important effect on protein kinase C activity and location (Nishizuka, 1992). This is a classic example of an inactive (precursor) molecule being converted to a biologically active product. This reaction is an excellent example of the role of phospholipids in membrane processes through an enzyme-catalyzed reaction.

Platelet Activating Factor

Platelet activating factor is a potent phospholipid agonist. Until the late 1970s, the concept of an intact phospholipid acting as an agonist was not a topic of great interest since it seemed to be such a remote possibility. However, in 1979 the finding that a compound associated with the development of systemic anaphylaxis in IgE rabbits was a phospholipid changed our views in a significant manner (Pinckard et al., 1979; Demopoulos et al., 1979; Hanahan et al., 1980). The structure proof on the naturally occurring material and comparison with a chemically synthesized form showed it to have the structure shown in Figure 1-13.

Two particular unique features of this molecule are the presence of an alkyl (OCH_2R) residue instead of an acyl (OCOR) group at the *sn*-1 position and the presence of a short-chain acyl group (i.e., the acetyl moiety) at the *sn*-2 position. The latter finding was quite remarkable since fatty acyl groups with less than 14 carbons are not found in mammalian phosphoglycerides.

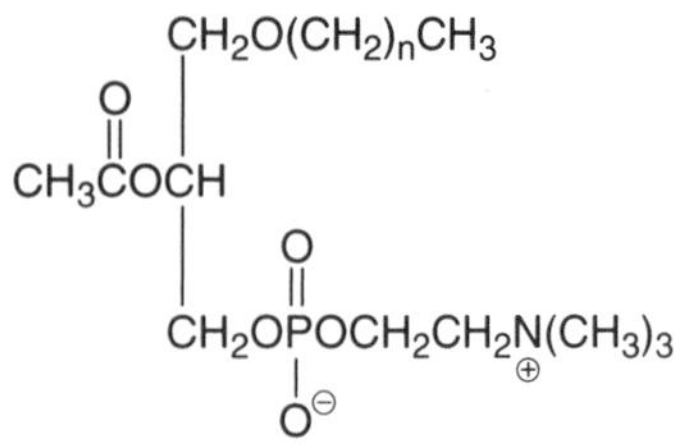

$$CH_2O(CH_2)_nCH_3$$
$$\underset{\underset{\displaystyle \|}{O}}{CH_3CO}CH$$
$$CH_2OPOCH_2CH_2\overset{\oplus}{N}(CH_3)_3$$
$$O^{\ominus}$$

Platelet activating factor
(1-O-alkyl-2-acetyl-sn-glycero-3-phosphocholine; n=15 or 17)

FIGURE 1-13. Platelet activating factor, a novel phosphoglyceride with potent agonist activity.

Platelet activating factor is a very potent agonist with an EC_{50} value near 1 $\times$ 10^{-10} M on rabbit platelets (aggregation, secretion) and an EC_{50} near 1 $\times$ 10^{-9} M for human platelet (aggregation) activation. As with many other agonists, platelet activating factor stimulates the activation of cellular phospholipases and initiates the turnover of several membrane phospholipids and the phosphorylation of specific proteins. Several reviews are available that chronicle the history and current status of this field (Snyder, 1982; Hanahan, 1986; Braquet et al., 1987).

Recapitulation: The Take-Home Message

The object of this chapter was to provide some insights into the exciting area of phospholipid chemistry and its relation to the *in vivo,* biological milieu. These fascinating, and at times challenging, compounds are important components of all mammalian systems from the isolated cell to the total organ and the fluids bathing the organ and cells. Simply on a mass basis they must be considered as intimate facets of cellular function. Their ubiquitous and multifaceted structural characteristics at once define the problem and at the same time the challenge. How can we hope to solve the mysteries of the behavior of these compounds, especially since there are well over 1000 different species of phospholipids in the mammals? Certainly one obvious approach is to isolate these molecules from a cell, prove their chemical structure, and then armed with this critical information progress to the next stage in which their role in the biological domain would be elucidated.

As a first approach, one must undertake to isolate these substances from a cell, purify them to apparent homogeneity, and then determine their chemical structure. Only with such an approach can one hope to interpret their exact role in cellular reactions. The next chapter will center on the isolation, purification, and initial characterization of cellular phospholipids.

TWO

PHOSPHOLIPID(S) ASSOCIATIONS IN CELLULAR STRUCTURES

Introduction

In the preceding chapter, the intent was to provide the reader with a broad perspective on the chemical characteristics of cellular phospholipids. At the same time, emphasis was placed on the potential usefulness of this information in dissecting the importance of phospholipids in cellular events, such as signal transduction. There is no doubt that the large number of observations reported on the close relationship of phospholipids to the transduction process has stimulated a widespread (and gratifying) interest in these compounds. Certainly it is very clear now that stimulus-induced activation of cells leads to the turnover of specific membrane phospholipids. The following diagram reemphasizes several, but not all, possible reaction pathways that can be invoked during an agonist (stimulus)–induced activation of a cell and gives the possible sequelae:

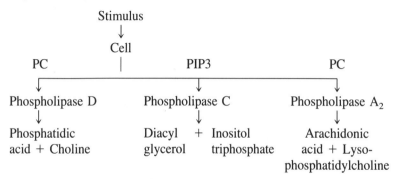

In each of the above reactions, the substrates phosphatidylcholine and phosphatidylinositol bisphosphate normally are considered biologically inactive in membranes. Then, subsequent to activation of cellular phospholipases by a stimulus, biologically active products are formed from these compounds. Thus, inositol bisphosphate triggers the release of calcium ions from intracellular stores, diacylglycerol is implicated in the translocation and activation of protein kinase C, arachidonic acid can be converted to biologically active prostaglandins, and phosphatidic acid can be an agonist in its own right. The major point to be stressed here is that phospholipid turnover is intimately associated with the signal transduction pathway in cells. Hence an understanding of the chemistry of these phospholipids is of major relevance to delineating the complicated process of signal transduction. While investigation of the behavior of phospholipids in this pathway in platelets has been a consuming interest of this author, the main thrust in this book will be simply to acquaint the reader with the chemistry of phospholipids of major importance in signal transduction and also to discuss other phospholipids found in mammalian membranes. Inasmuch as most investigations on stimulus response in cells utilize quite small numbers of cells—for example, a typical experiment on human platelets might use 1×10^9 cells, which would yield ~50 μg of total lipid—this poses a challenge to an investigator to be able to isolate and identify these lipids. This will be the general direction at this point. However, liberal use will be made of information gathered on the structure of cellular lipids through use of large amounts of cells or organ tissue. This latter approach has been of great advantage to structure proof studies on individual phospholipid fractions and is of great benefit in identification of the small amounts of phospholipids available in the usual signal transduction experiment.

Among the key events that have triggered the enormous scientific interest in naturally occurring lipids over the past 20 years have been the impressive strides made in establishing methods for their isolation from biological materials, their purification, and their structure proof. Certainly, advances such as the chemical synthetic routes to reference compounds for comparative structure proof and identification have been of paramount importance. In addition, one must consider column chromatography, thin-layer chromatography, infrared spectrometry, gas-liquid chromatography, and mass spectrometry as techniques and procedures that have revolutionized our capabilities in this area. Notwithstanding these considerable advances, the isolation and purification of a phospholipid in particular still represents an adventure fraught with difficulties unless the investigator exercises considerable care and is aware of the chemical nature of the system under investigation and of the possibility of artifact production.

Noncovalent Binding of Lipid to Protein

Current evidence strongly supports the presence of two classes of binding between lipids and proteins in tissues. The first, which represents nearly 98%

of the lipid, is the noncovalent interaction. The second class, which corresponds to 2% or less of the cells lipid, is characterized by covalent linkage of lipid to protein. A modest discussion of these two classes of binding is presented next.

Certainly the fact that neutral solvent mixtures (e.g., methanol and chloroform) are capable of extracting the major amount of tissue lipid at room temperature supports a noncovalent type of linkage. Even freezing of a tissue and allowing it to warm to 4°C in diethyl ether can allow a good recovery of lipid. Armed with these observations, it is of value to reflect on possible types of interactions which could provide a noncovalent association between lipids and proteins in cellular membranes and other tissues. These interactions can involve polar groups (such as esters), charged groups (such as phosphatidylserine and histidine residues in proteins), and unsaturated moieties (such as oleic acid and phenylalanine). Among the nonpolar groups would be (a) the hydrocarbon chain of a fatty acid such as palmitic acid and (b) the hydrocarbon chain in valine. These forces are described to a brief extent in the examples that follow.

Electrostatic Forces (Charged Groups)

These are essentially forces operating through mutual Coulombic attraction or repulsion of the net charge or electric moment carried by two interacting molecules. These forces undoubtedly are of importance since many lipids and proteins bear a net charge at physiological pH. In the example cited below, the possibility of interaction of the charged groups on phosphatidylethanolamine with charged groups on a peptide is illustrated as a partial formula in Figure 2-1.

In this instance the interaction of the positively charged amino group of phosphatidylethanolamine and a negative charged carboxyl group of a peptide (aspartic acid) would account for a binding point of a lipid to a protein molecule. The interaction or attraction energy for this type of binding may be near 5 kcal/mol at a relatively narrow distance of 5 Å. Furthermore, the negatively charged phosphate (oxygen) group can attract a positively charged group on a peptide chain, particularly a lysine residue, and hence within a short distance a double interaction can occur. These interactions can be repeated many times over in lipid-protein complexes and in general can be considered operative to a distance of 10–20 Å.

Polarization Forces

These very weak forces arise by virtue of the polarization of a molecule by the charge or permanent electric moment on an adjacent molecule. In addition, polar bonds may result in bonds between atoms of greatly different electronegativity in which the shared pair of electrons are distorted from a symmetrical distribution. These "induction" forces, though of low energy, are of some import in the binding of lipids and proteins.

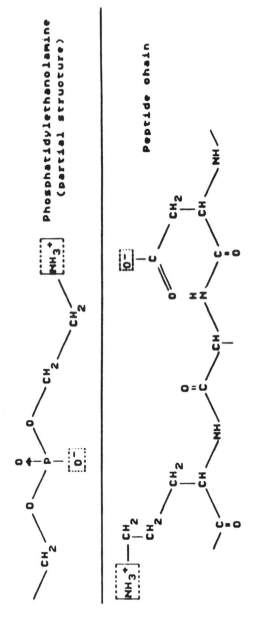

FIGURE 2-1. An illustration of the interaction via polarization forces between membrane phospholipid and peptide.

As an example, the charged phosphate group on phosphatidylethanol-amine, for example, can interact with the hydrocarbon (CH_2) chain of an amino acid—for example, valine—in a peptide. A similar situation would hold in the example to the right for interaction of the hydrocarbon unit in a peptide chain. In both instances the groups with permanent dipole moments can induce a temporary dipole moment in an adjacent molecule. These inter-actions, however, are very weak and act only at very short distances; thus the polarization energies may be of the order of 0.002–0.004 kcal/mol at a distance of 5 Å.

London–van der Waals Dispersion Forces

The interaction of two neutral, nonpolar molecules such as two hydrocarbon chains at short distances produce London–van der Waals dispersion forces. Certain requirements of this interaction are that the molecules are in their ground electronic state and that no electron exchange occurs between the two molecules. In a nonpolar molecule the shared pairs of electrons are rather symmetrically distributed and hence no decided regions of positive or nega-tive charges occur. Consequently weak attractive forces are in operation and charge density fluctuations can induce an instantaneous electric moment on one molecule, which then can induce an electric moment on another mole-cule. These (average) interactions then constitute the London–van der Waals forces which are thought to occur in lipid-protein complexes. Although they represent an energy of low magnitude (~0.1 kcal/mol at 5 Å), these forces are of importance in complexes where large numbers of saturated hydrocarbon units are present, and the repetitive nature of these groups (large number of combining sites) make the London–van der Waals forces of great importance in holding or stabilizing a lipoprotein.

A similar attractive force is noted in the bonding of a hydrocarbon unit of a fatty acid to the hydrocarbon unit of certain amino acids, in the bonding of the hydrocarbon units of fatty acids to the nucleus of cholesterol, and so on.

Hydrogen Bonding

This type of bonding, which is of considerable importance in maintenance of the structural integrity of proteins, also plays a role, although perhaps not as important as the one just described, in association of lipids with proteins. It is not within the province of this present discussion, however, to consider this type of interaction in detail.

A Summation

Present evidence shows that the association of lipids with proteins can be accomplished through attractive forces, which on the basis of a single unit are of low energy magnitude, but on the basis of multiple repeating units present the necessary cohesive forces for these complexes. At the same time, these forces are instrumental in the association of lipid with lipid, which also occurs in these complexes.

The structure of the soluble lipoproteins has been examined in great detail, and the general consensus is that the apoproteins (lipid-free) possess a high level of α-helical structure when associated with lipid. Thus, the amino acid sequence of these helical regions have an ionic side chain at every third or fourth amino acid. A current view of the structure is that a polar region or edge is present within the longitudinal axis of the helix, and this interacts with the phospholipid polar head group and the aqueous media on the outside. The opposite side of the helix is very hydrophobic in nature (due to side-chain groups), and these associate with the apolar core of the phospholipid molecule.

A similar type of interaction is thought to occur in membrane lipoprotein molecules. The problem in the latter studies is that the membrane apoproteins are not easily solubilized. If further information on the structure of biological membranes is required, then it is recommended that a recent book by Petty (1993) and one edited by Wirtz et al. (1993) be placed on a "must" reading list. An older, but very good, short review on lipid-protein interaction possibilities in membranes is one presented by Danielli (1982), who is widely recognized as a pioneer as well as a legend in this field.

Covalent Binding of Lipids to Membrane Proteins

While there is no doubt that the major attractive forces maintaining lipids in membranes are noncovalent in nature, there is excellent evidence in the literature showing that a small percentage of the membrane lipids is covalently linked to membrane proteins. These lipids are highly specific in nature and will be discussed briefly below. Normally these lipid protein complexes are not found in the organic solvent phase of a typical (lipid) extraction procedure. Rather they would be found in the water-rich phase of such an extraction approach. Basically there are four specific classes of lipid covalent binding to protein:

1. Palmitic acid attachment via a thioester linkage as depicted in Figure 2-2
2. Myristic acid binding resulting in an amide linkage as shown in an abbreviated form in Figure 2-3
3. Glycosylphosphatidylinositol interaction via a carboxyl terminus grouping as described in Figure 2-4
4. Prenylation of proteins by 15- and 20-carbon-chain prenyl groups presented in partial formulation in Figure 2-5

These various derivatives appear to be very important components of regulatory systems in the cell. For further consideration of this fascinating and

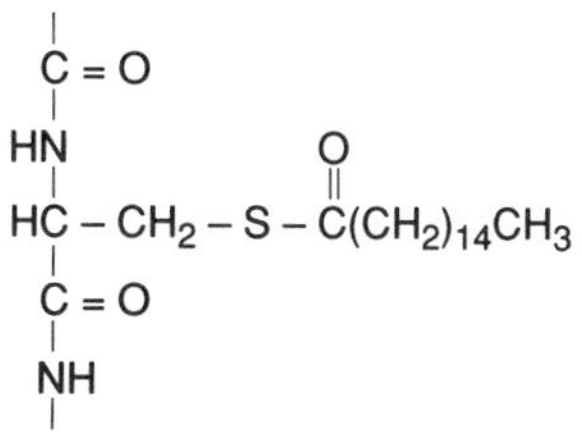

$$\begin{array}{l}
\overset{|}{C} = O \\
\overset{|}{H}N \qquad\qquad O \\
\overset{|}{HC} - CH_2 - S - \overset{\parallel}{C}(CH_2)_{14}CH_3 \\
\overset{|}{C} = O \\
\overset{|}{N}H \\
|
\end{array}$$

FIGURE 2-2. The partial chemical structure of palmitic acid bonding to a thiol group on a cellular peptide.

important topic, excellent articles are available by Towler et al. (1988) and Stimmel et al. (1990).

Recovery of Phospholipid from Cellular Membrane Structure

General Comments

In the association of phospholipids with proteins as described earlier, the major type of interaction appears not to involve any covalent linkages. Rather, these combinations in biological materials are attributable to the interaction between comparable functional units in these two groups of compounds—for example, the apolar, hydrophobic residues of long-chain fatty acid found in phospholipids—and comparable amino acid residues of proteins. Additional associations can derive from polar and charged groups, such as the carboxyl or phosphate residues. Also, steric fit must play a role and water must be considered as an important participant. The ability of lipids (such as phospholipids) to swell in water and for proteins to bind definitive quantities of water has been well established for many years. Thus, water must be considered as an important constituent of any lipid-protein complex, whether it is a soluble plasma constituent or an insoluble component found in membranes. This point is well illustrated by the fact that extracts of phospholipid from any biological sample require the presence of a polar solvent, such as methanol, plus ultimately a nonpolar solvent, such as chloroform, for complete extraction. Simple addition of diethyl ether to a tissue such as minced liver will allow extraction of vanishingly small amounts of lipid. However, if this

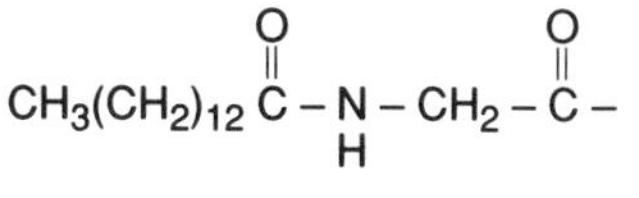

$$CH_3(CH_2)_{12} \overset{\overset{\displaystyle O}{\parallel}}{C} - \underset{\underset{\displaystyle H}{|}}{N} - CH_2 - \overset{\overset{\displaystyle O}{\parallel}}{C} -$$

FIGURE 2-3. An abbreviated formulation of the amidation of an amino acid group (on a peptide) by myristic acid.

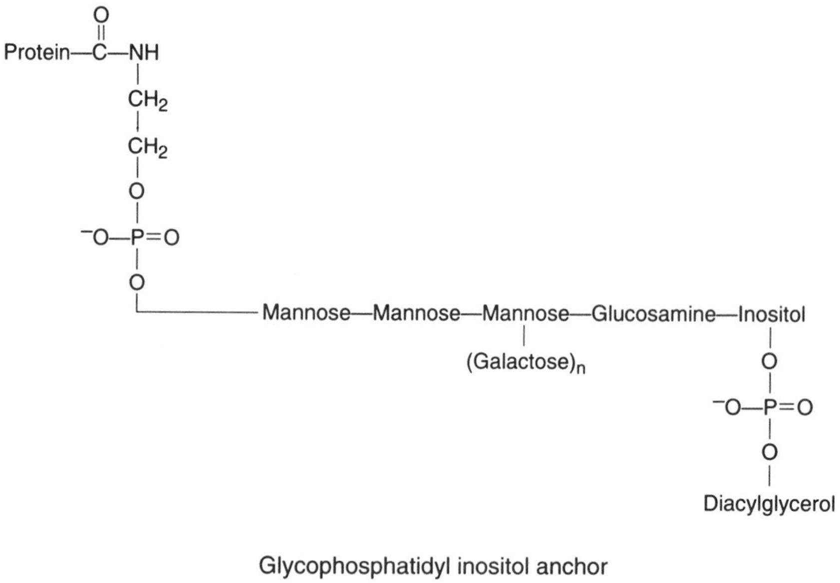

$$O$$
$$\parallel$$
Protein—C—NH
|
CH$_2$
|
CH$_2$
|
O
|
$^-$O—P=O
|
O
|_____ Mannose—Mannose—Mannose—Glucosamine—Inositol
| |
(Galactose)$_n$ O
|
$^-$O—P=O
|
O
|
Diacylglycerol

Glycophosphatidyl inositol anchor

FIGURE 2-4. A partial chemical structure of the glycosyl PI anchor of mammalian cells.

sample had been frozen first, then allowed to melt in the presence of diethyl ether, significant quantities of lipid could have been removed. Obvious disruption of an "ordered" water structure allowed access of this relatively apolar solvent to the lipid.

Even though our understanding of the possible types of lipid-protein interactions in membranes has developed only recently, it was evident to early investigators that neutral solvents per se were the most effective for isolation purposes. Perhaps the most widely used solvent extraction procedure for many years was that employing a mixture of ethanol–diethyl ether (usually 1:3, v/v). This technique involved extraction of a tissue with this solvent combination for several hours at 55–60°C (Bloor, 1928). However, as more refined techniques were developed for the detection and assay of lipids, it became evident that this solvent (and condition) could have a deleterious

FIGURE 2-5. Protein modification by prenylation in cells.

effect on certain lipids. A case in point concerns the activation of certain lipolytic enzymes by this solvent. It had been reported that cabbage leaves had an unusually high concentration of phosphatidic acid, and it was assumed that this phospholipid was an important metabolite of this tissue. However, the high level of this particular phospholipid in cabbage leaves lipids was shown to be artifactual, since the combination of this particular solvent system and an elevated temperature served only to promote high activity of a specific (and the newly discovered) phospholipase D (Hanahan and Chaikoff, 1947). This enzyme catalyzed the cleavage of nitrogen bases (such as choline and ethanolamine) from the parent phosphoglycerides to yield phosphatidic acid. However, it should be noted that at the time of the report in the 1940s on phosphatidic acid in cabbage, enzymology was still a field in its infancy, and certainly any proposal that an enzyme could be activated by an organic solvent would have met with much disfavor or more likely disbelief. Notwithstanding the historical facets of these observations, ethanol–ether mixtures have been replaced by a much more efficient and reliable system.

Undoubtedly, the most widely employed solvent system in use today is that of chloroform and methanol. Essentially this solvent mixture was developed by Folch et al. (1957) and employed chloroform and methanol in a ratio of 2:1 (v/v). As initially described, a tissue-solvent homogenate was filtered and the filtrate washed with a large volume ratio of water. Though the extraction per se was highly efficient, could be conducted at room temperature (or lower), and was rapid, the intractable emulsions which frequently resulted during the washing procedure (for removal of nonlipid contaminants) posed serious and certainly exasperating problems. Among the latter were loss of more polar lipids (e.g., inositides, sialoglycolipids) and an inordinate long time and effort required to break the emulsions. Folch later modified the wash routine to prevent emulsions and loss of lipid through inclusion of salt in media and use of a solvent wash comparable in composition to that of the upper phase (i.e., predominately a methanol–water mixture). Though the Folch procedure served admirably for some period of time, it has been largely superseded by a modification described next.

Currently, there is no doubt that the most widely used method for extraction of tissue lipids is that of Bligh and Dyer (1959). Basically, this is a modification of the Folch method and employs a careful calculation of the amount of sample (tissue) water such that the overall mixture will have a final composition of chloroform–methanol–water of 1:2:0.8 (v/v). Thus, a single-phase extract can be obtained and extraction completed very rapidly, even within minutes. Recovery of the lipid in a chloroform-rich phase can be achieved by addition of equal volumes of chloroform (under certain conditions) and water to produce a two-phase system. The lower ($CHCl_3$) phase is subsequently washed with a methanol–water (1:0.9, v/v) mixture to allow removal of a substantial amount of the nonlipid contaminant with little or no problems with interfacial "fluff" formation or emulsions. However, even though this is a highly efficient method, it is still advisable that one take steps to remove all nonlipid contaminants by the techniques described in the next section. Inherent in all these operations is the attention of the investigator to such mundane but highly important points as purity of solvents (chloroform readily forms hydrochloric acid; methanol can contain aldehyde), prevention of undue exposure of sensitive fatty acids (i.e., polyunsaturated types) to air and light exposure, and proper storage conditions (usually in a solvent containing as high an amount of methanol as possible).

Extraction Procedures

The isolation of lipids from cells or tissues is not as simple and straightforward as one might desire, but is essentially an important adjunct to characterization of membranes (composition, lipid-to-protein ratio, structure proof, definition, new lipids, etc.). While this is recognized by many investigators in the field, it is difficult for the novice in this area to become aware of some of the potential problems in extraction procedures and the reasons for particular approaches. Thus it seems fitting at this point in time to comment on some of the nuances of the approaches used in isolation, purification, and identification of lipids present in cell membranes. These topics are subdivided into areas which are considered to be of major import to a successful consideration of the extraction procedure.

Solvents

There are overwhelming data to support the conclusion that the major part of any cell or tissue lipid is present as a lipid-protein complex. Primarily, as mentioned earlier it appears to be a bonding involving water molecules, principally through hydrogen bonding, though aliphatic hydrocarbons can play a role. The support for noncovalent interaction and the role of water is that neutral solvents such as ethanol or methanol are very effective in breaking this interaction of lipid to protein. However, if one chooses to use diethyl ether alone or chloroform alone, this will allow a poor recovery of lipid from cells or tissue, at least at room temperature. However, if the tissue is frozen first and then these solvents added during the thawing, it could allow a reasonably good extraction. It is not a particularly advantageous route but it illustrates the point. In general, one includes a polar neutral solvent such as methanol and a non-polar solvent such as chloroform in most extraction of lipids.

Temperature Effects

Ethanol–Ether Extraction

It has been only a relatively few years ago that the solvent of choice was a combination of ethanol and diethyl ether (3:1) used at reflux temperatures. This was commonly referred to as *Bloor's extraction solvent* (Bloor, 1928). It was effective in many instances but could cause the following deletorious effects on lipids.

YEAST. An older, but still applicable, observation involves extraction of baker's yeast at 55–60°C in ethanol–diethyl ether (1:3, v/v). The phospholipids are dramatically affected by this procedure. They are obtained as dark viscous lipids obviously peroxidized with high N:P ratios (6–8:1). However, the same yeast was extracted at 25°C in the same solvent, the isolated lipids are light tan in color and not easily oxidized, and the individual phospholipids have N:P ratios near 1, as expected. This example does give a clue as to the deleterious effects of high temperature.

PLANT MATERIAL. A particularly dramatic example of potential alterations is shown with ethanol–ether extraction of plants, such as cabbage or carrots at reflux temperature. Under these conditions the enzymatic attack of phospholipids is accelerated. Hence, reports of high levels of phosphatidic acid in these and other plant lipids were directly attributable to activation on phospholipase D. It is now well established that this enzyme as well as other phospholipases are stimulated, activated, or active in organic solvents. Further details and complete elucidation of these effects will be presented in Chapter 4. Certainly this illustrates that lower temperature and the use of other solvents will avoid the above type of degradation.

Oxidative Alterations

A common problem that can emerge during the isolation and handling of phospholipids (as well as other lipids containing unsaturated linkages) is that of oxygen attack at the double bonds in the long-chain fatty acids. This process is often called *lipid peroxidation.*

Oxygen attack on the methylene-interrupted diene system in fatty acyl residues on phospholipids can be summarized by the abbreviated equations in Figure 2-6.

In the reaction shown in Figure 2-6, the abstraction of the H from structure 1 can be initiated in a number of ways with substances such as free metal ions, hematin (a metabolite of hemoglobin, basically a hydroxide of the trivalent iron derivative of heme), or cytochromes. There is an equilibrium between the free radical intermediate (a *cis–cis* free radical, structure 2 and the rearranged *trans, cis*-conjugated diene (structure 3). The addition of oxygen to structure 3 obviously would have a significant effect on this equilibrium, leading to increased amounts of the peroxyl derivative (structure 4). This latter derivative then can accept a hydrogen atom from another *cis, cis*-unsaturated fatty acid (such as linoleic acid) to perpetuate the reaction initiated with structure 1. Although one form is shown for structure 2, there actually would be a total of three resonance hybrid free radicals. Finally, antioxidants are thought to act by preventing hydrogen addition to the derivative (structure 4), thus preventing continuation of the chain reaction.

Many investigators add an antioxidant, such as butylated hydroxytoluene

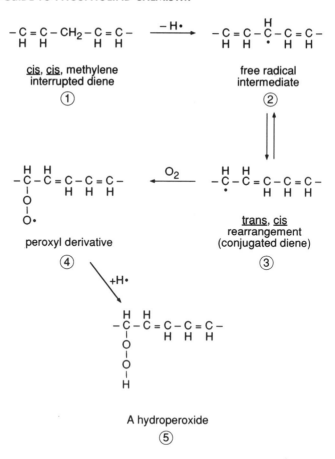

FIGURE 2-6. The chemical sequence in peroxidation of unsaturated fatty acids (usually polyunsaturated) in membrane phospholipids.

(BHT), to a crude lipid extraction to prevent oxidative attack in the olefinic (unsaturated) components. If lipid extracts are to be stored for any length of time (certainly always at low temperatures), then it is very wise to include such an antioxidant. An alternative approach, which is not used that often, is to store the lipid extract in a dried form under oxygen-free nitrogen in a sealed glass tube. In the author's experience, if the lipid mixture is subjected to immediate chromatographic separation into the individual phospholipid components, the chances for oxidative degradation are reduced to a significant degree. This is attributable, at least in part, to the removal of pigmented substances (nonphospholipid) which can function as catalysts for oxidative changes in individual phospholipids. However, if one is working with phospholipids with a known high level of polyunsaturated fatty acids, it is good practice to include an antioxidant such as BHT.

If further information is desired on the peroxidative process, an excellent review by Halliwell and Chirico (1993) is recommended reading.

Contaminants

In any of the lipid isolation procedures described above, there is always the potential for inclusion of "nonlipid" components in the total lipid extract. These contaminants can include free amino acids, peptides, pigments, carbohydrates, inorganic salts, and other compounds.

While there is no doubt that one must remove these nonlipid substances so that highly purified lipid may be obtained for particular study, it is interesting to reflect on another aspect of the problem. On the one hand, it is entirely conceivable that the nonlipid components are simply adventitious contaminants and bear no relation to the lipid behavior as such. On the other hand, there is developing evidence that substances, such as divalent cations as well as carbohdyrates, may have close associations with lipids of a tissue, and these interactions might have an important metabolic consequences. Phospholipid complexes of import to the transport of certain amino acids across the intestine have also been reported. Certainly the association of divalent cations (i.e., magnesium or calcium) with the membrane could play a decisive role in the behavior of a particular phospholipid, most likely the acidic types such as phosphatidyl serine or phosphatidyl inositol. Thus it is evident that careful consideration must be given to the type of complexation which may be of great importance to the stability and behavior of a membrane system.

Notwithstanding the possible association of the aforementioned nonlipid components with lipids in tissue, the fact remains that if the objective is to obtain a clean, highly purified preparation of phospholipids for further chemical study, for example, then these substances must be removed.

One of the more widely used techniques for removal of nonlipid contaminants is that first described by Folch et al. (1951). Essentially in this wash procedure, a chloroform–methanol solution of lipid is mixed with an aqueous salt solution in a ratio calculated to give two phases: The upper, water-soluble phase would contain the contaminants, while the lower phase would contain the desired lipid components. This was a useful procedure at the time and worked quite well for removal of most, if not all, impurities. Perhaps its greatest drawback centered on the formation of rather intractable emulsions which often prove frustrating to handle and also lead to losses of lipids. Though the Bligh-Dyer extraction procedure has proven effective in providing quite pure lipid samples, it is suggested that additional treatment, namely the use of Sephadex G-25 column, be employed for the removal of impurities. This technique was introduced by Wells and Dittmer (1963) and can be used as a "de rigeur" procedure for initial purification of lipid extract. Essentially it involves the passage of a lipid extract (chloroform–methanol–water) preparation through a Sephadex G-25 column. The lipid is excluded from the Sephadex and rapidly passes through, whereas the water-soluble components (contaminants) are included in the beads. However, one must be cognizant of the point that none of the purification schemes are foolproof. The Sephadex treatment just described causes retention of gangliosides and can cause some problems. This is clearly indicated in a study by Carter and Kanfer (1973) in which the means by which nucleotide donor sugars can be removed from gangliosides was explored. This was deemed of import especially in experiments in which radioactive nucleotide donor sugars were used in an *in vitro* investigation of the biosynthesis of gangliosides from other lipids, but not from the donor sugars. Also the Sephadex G-25 procedure as described by Wells and Dittmer was most effective in accomplishing one of the desired goals (namely, the removal of the sugar nucleotides) but failed in another (that is, some 25% of the applied ganglioside was apparently trapped by the gel). These authors finally devised a procedure in which a Folch-type chloroform–methanol partitioning was found most effective and where, through the use of calcium at low pH, the gangliosides are partitioned quantitatively into the chloroform fraction. Essentially this demonstrates that no one purification scheme is best for all of these lipids and that on the strength of this basic premise, one must always carefully evaluate quantitatively the nature of the lipids in the various preparations after specific treatments.

A Transition: Facing the Identity Crisis for Phospholipids

The major challenge in any study in which phospholipids are central components in a biological reaction of interest or in which there is evidence for a new type of phospholipid is their identification. While there are many excellent techniques available to accomplish this goal, there is absolutely no magic route to a satisfactory identification of specific phospholipids in a crude lipid extract except through the use of separation procedures. Basically it is next to impossible to identify or detect a specific phospholipid species in such a complex mixture. Thus, one must bite the bullet and resort to several analytical techniques for definitive proof. Perhaps the most important facet of this approach is that the compounds must be extracted from a tissue and then subjected to chromatographic separation.

In an extraction procedure using chloroform–methanol mixtures, all of the noncovalently bound lipid is extracted. By far the most popular and convenient method for separation of a complex lipid mixture is that of thin-layer chromatography. While attention in this book will be centered only on the phospholipids found in mammalian tissue extracts, this is not to downplay or diminish the importance of a group of non-phosphorus-containing lipids found in the usual extract, namely the polar glycolipids. Essentially these are the gangliosides and the neutral sphingoglycolipids. Structurally all of these compounds contain sphingosine as the lipid backbone. Though mention will be made in the next chapter of their detection on thin-layer chromatograms, the reader is urged to consult the excellent chapter by Hakomori (1983), in which he describes the chemistry and isolation of these unique compounds.

THREE

ISOLATION AND IDENTIFICATION OF CELLULAR PHOSPHOLIPIDS

Introduction

In this chapter, some of the types of methodologies currently in use for isolation and analysis of cellular phospholipids will be outlined. Such techniques can be readily applied to experiments designed to explore the involvement of phospholipids in cellular events, such as stimulus-induced activation. Primary attention will be paid to the human platelet. If you need to justify the choice of human platelets as the cell of choice, a number of highly creditable reasons can be cited. Only three need to be considered at this point. First, the circulating platelet is of paramount importance in hemostasis, and there is convincing evidence that its membrane phospholipids are intimately involved in this process. Second, these cells can serve as excellent targets or model systems for stimulus-induced activation in which the membrane phospholipids play an important role. Third, human platelets can be isolated from whole blood by a simple, convenient centrifugal approach. Human donors are available at a very reasonable cost, and the platelets obtained from a varied spectrum of donors show remarkable consistency. Thus, one can undertake their isolation using the following method and have them available for immediate experimentation.

Isolation of Human Platelets

Nonfasting venous blood is drawn (with informed consent) from male or female subjects between the ages of 20 and 40 years, who are considered to be normal and healthy and had not ingested platelet-active medication for at least 10 days prior.

Blood is obtained by insertion of a butterfly infusion set (12-in. tubing from Abbott Hospitals, Inc., North Chicago, IL) with a 1-in. × 19-gauge needle into the antecubital vein. A few milliliters of blood is allowed to flow before a 60-ml plastic syringe containing 7.5 ml of an ACD [10.8% citric acid, 2.2% trisodium citrate, and 2% dextrose (w/v)] solution was attached. Four syringes were filled to the 50-ml mark and inverted gently, and the contents were transferred carefully into 50-ml plastic tubes that are then capped. The blood-to-ACD volume ratio is 6:1 (v/v). The tubes are centrifuged at 2000 rpm (830g) in a Sorvall RT 6000 centrifuge at 24°C for 15 min. The platelet-rich plasma (PRP) is carefully transferred into 15-ml plastic conical centrifuge tubes using a plastic syringe with a plastic tip extension (Fibro-tips from VWR Scientific, Houston, TX), leaving a safe margin of plasma above the buffy coat layer. Adenosine (~10–20 μM, final concentration) is added to ~13 ml of PRP in each tube. Tubes are capped, inverted gently, and centrifuged for 20 min at 2200 rpm (1000g) at room temperature. The supernatant is decanted, and the pellets are quickly but gently resuspended in a few milliliters of wash buffer by using a Pasteur pipette to suction the buffer back and forth into the pellet. More wash buffer is added to make a total volume of about 13 ml, and again adenosine (~10–20 μM, final concentration) is added. The tubes are capped and gently inverted and then centrifuged for 10 min at 1800 rpm at room temperature.

After the supernatant is decanted, the pellet is resuspended as before in a limited amount of buffer, pH 6.5, without ACD and combined into one tube (a total of 3–4 ml). A Coulter Counter Model S plus IV is used to count a 1:10 dilution of platelets in buffer. The platelets are then adjusted with the final suspension buffer to a concentration of 1.25×10^9 cells per milliliter. These platelets, as small aliquots, are stored in capped tubes at room temperature. As needed, these aliquots are used for specific experiments.

The best way to test the viability of the platelets is to evaluate their response to a standard stimulus, such as thrombin. The biological endpoint, which is reached in a concentration dependent manner, is aggregation.

Aggregation Assay

Fibrinogen (10 μl of 20 mg/ml in saline) and pH 7.4 assay buffer with 1.3 mM Ca^{2+} (400 μl) are measured into a Chrono-log cuvette containing a magnetic stir bar and warmed at 37°C for 1–2 min. An aliquot of platelet suspension (100 μl) is added, and the cuvette is placed immediately into the stirring chamber (1000 rpm, 37°C) of a Chrono-log Platelet Aggregometer (Model 440, dual-channel unit) attached to a Chrono-log Model 706/7 Function Module Recorder. Within 1 min, 5 μl of ADP (usually 10 μM, final concentration) is added, followed at ~1 min by 5 μl of agonist (various concentrations). Platelet aggregation is determined at 2 min by the percentage of change in light transmission. As a control, 0.2 U/ml thrombin was used under similar conditions. Without the preaddition of ADP, the amount of aggregation caused by 0.2 U/ml thrombin was markedly variable from one preparation to another.

General Comments

Platelets isolated under the experimental conditions described here are remarkably consistent in their responses to stimuli. The normal laboratory life span for such a cell preparation is 6–7 hr at room temperature. This time frame is sufficiently broad enough to allow meaningful experiments to be done on the turnover of cellular phospholipids under a variety of experimental conditions.

In the usual instance the average yield of platelets from 200 ml freshly drawn venous blood (the usual volume from a single donor) is 8.5 ml, containing 1.25×10^9 platelets per milliliter. This translates into ~645 μg total lipid/10^9 cells.

The Next Great Step: Isolation of Platelet Lipids

It is assumed at this point that the lipid composition of the human platelets is unknown to you and that acquisition of such data is mandated prior to your planned biological experiments. The central theme will be to illustrate a very common and popular way in which to extract the lipids from cells, as well as to demonstrate how one would proceed to an initial inquiry into characterization of the species present.

General Procedure

The most widespread and effective solvent system for this purpose is the Bligh-Dyer procedure, which was mentioned earlier in Chapter 2. A typical protocol that can be employed for total lipid extraction follows.

The platelet preparation is mixed (depending on the volume of sample and the cell count) with sufficient methanol (usually added first) and chloroform to make a final mixture of chloroform–methanol–water (based on water in cell sample) of 1:2:0.8 (v/v). This mixture is then mixed well and allowed to stand for 25–30 min at room temperature in a dark cabinet (or shielded with aluminum foil) to allow extraction of the lipids from the cells. Then the mixture is centrifuged at 2000g for 10 min. A single-phase, clear-colored supernatant will result. This is carefully removed from the pellet and saved, because it represents the total lipid extract. Though a second extraction of the pellet with chloroform–methanol–water (1:2:0.8, v/v) can be done, it is usually not necessary.

The clear supernatant is then mixed with 0.5 volume chloroform and 0.5 volume water and vigorously mixed using a laboratory vortex unit. Phase separation occurs and usually a well-defined upper and lower phase is obtained. However, these phases may not always be clear and it is necessary to centrifuge this mixture at 2000g for 10 min.

Then the water-rich upper phase is carefully removed and saved. The chloroform-rich lower-phase, *which contains the desired phospholipids*, is washed twice with methanol–water (10:9, v/v) using 0.5 volume of this solvent. In each washing, the protocol is to vortex the mixture vigorously and then centrifuge to allow complete separation of the phases. The clear, water-rich upper phase is carefully removed and can be saved and mixed with the original water-rich upper phase.

In the author's experience, extensive washing with methanol–water can lead to a seemingly intractable emulsion, but this can be broken by centrifugation of the mixture at 2000g for 10 min at room temperature. The final chloroform-rich (lower) phase is evaporated under a nitrogen (oxygen-free) stream in a fume hood and then dissolved in chloroform–methanol (2:1, v/v). On occasion, this extract may be slightly turbid; if so, simply evaporate again to dryness under nitrogen and redissolve in the same solvent mixture as before. It should be made to volume in a tightly fitting glass-stoppered volumetric flask. This extract then essentially represents the total lipid of the cell. Realistically the lipid covalently bound to protein is not extracted by the above procedure. Nonetheless, for our purposes the total lipid extract noted here will be the center of our attention in this chapter. If the sample has been stored at low temperature, be sure to allow it to warm to room temperature and to mix it well since solvent and sample layering can occur.

Next, the removal of any suspected contaminants—such as carbohydrates (e.g., glucose), free amino acids, nucleotides and so on—can be accomplished by the procedure of Wells and Dittmer (1963). The lipid sample, dissolved in a mixture of chloroform–methanol–water (60:30:4.5, v/v), is passed through a previously washed column of Sephadex G-25, the effluent is collected and saved, and the column is then washed with a mixture of chloroform–methanol (2:1, v/v). The second eluent is collected and combined with the first and will contain all the phospholipid, free of contaminants. The two eluates are combined, phased by the addition of water, and then the chloroform-rich layer is removed and evaporated to dryness under nitrogen. The residue is dissolved in chloroform–methanol (2:1, v/v) and made to volume in a glass-stoppered volumetric flask.

This same extraction technique can be applied to erythrocytes, neutrophils, and other comparable cells. A similar approach also can be used with tissues (such as liver and spleen) employing 7–8 ml chloroform–methanol–water per gram of homogenized tissue. In the latter case, the extraction period should be a minimum of one hour at room temperature.

Storage and Handling of Lipid Extracts

An important facet of any investigation on the characteristics and composition of a lipid extract is in its handling and storage. If butylated hydroxy toluene

(BHT) is to be added as an antioxidant, a final concentration of 0.1% should suffice. As described previously, chloroform–methanol is the solvent of choice in most instances and if the sample is not to be worked up immediately, it should be stored in an explosion-proof freezer at −20°C. When this sample is brought to room temperature for analysis, be sure to mix well since these solvents (and sample) can undergo layering at low temperatures.

What To Do Next

There are several different analytical routes one can take at this point to characterize the lipid sample. It simply depends on how sophisticated an approach and how much information is needed. The following procedures will provide some qualitative insights into the components present in this extract and are adequate for an initial foray.

Qualitative Test for Lipid P

This simple colorimetric test will show whether phosphorus-containing lipids (i.e., phospholipids) are present, but will *not* prove the structure or composition of any of them. The test is conducted as follows:

Assuming that approximately 20 μg of lipid P has been extracted (from 1.25×10^9 platelets in this case), an aliquot containing 0.2–0.3 μg P is pipetted carefully onto a precoated silica gel G plate [measuring 2.5 cm × 10 cm (Analtech)] and allowed to air dry. The plate is then sprayed (in a fume hood) with the phosphorus stain reagent developed by Dittmer and Lester (1964). It is composed of a mixture of molybdenum oxide, MoO_3, in sulfuric acid and powdered molybdenum. As little as 5 μg of phospholipid (by weight) can be detected. Within two to three minutes at the most a bright blue color should appear. This technique will allow detection of phospholipid at levels as low as 5–10 μg (by total weight) or 0.2–0.4 μg (as phospholipid P). This calculation is based on the assumption that the average P content of a diacylphosphoglyceride is approximately 4%. Spray (glass) bottles of different styles and sizes are widely available from commercial glass equipment suppliers and from companies specializing in chromatography supplies.

Although this is a very simple procedure, the result will provide a basis for calculating the amount of sample needed for a quantitative P analysis as well as the amount needed for thin-layer chromatography.

The Next Appropriate Move

A particularly appropriate maneuver at this point is to obtain information on the general type of compounds present in the sample. This can be achieved quite easily by subjecting a sample to thin-layer chromatography. There are a number of different types of precoated thin-layer plates on the market now, and again it is simply a choice of what is to be accomplished. The same can be said of the solvents to be used: acidic, basic, or neutral; polar, apolar, or a mix. It would be to the advantage of a new investigator to consult a reference book on this subject. A recommendation is a treatise by Touchstone (1992). Prior to undertaking thin layer chromatography it would be comforting to know the actual amount of lipid P present and hence a quantitative P analysis is in order. The following is a description of this method.

Quantitative Phosphorus Assay

Currently the only satisfactory route to quantitative determination of total lipid phosphorus is through digestion of the sample to yield inorganic phosphate. Subsequent treatment of the digest with a reagent produces a colored product. A satisfactory digestion reagent is 70% $HClO_4$ and a suitable color reagent is aminonaptholsulfonic acid (ANSA)—though there are others in use. A blue color develops as a result of the interaction of inorganic phosphate and the ANSA. It can be used for quantitation as described in a modification of the procedure developed by Bartlett (1959). If the sample is suspected of containing any phosphonate bonds—that is, C–P–O instead of the usual diester structure, C–O–P—then it is necessary to alter the digestion conditions (Berger et al., 1972).

Experimental Procedure

Basically, a lipid sample containing lipid P in the range of 1–5 μg is transferred into an acid-washed Pyrex digestion tube and the solvent is removed. Be certain all the solvent is removed since perchloric acid can react quite violently with organic solvents. To the dried sample is added 70% reagent grade perchloric acid, and the tube is placed in a sand bath maintained near 200°C for 2 hr. The assay should be run in triplicate, and a set of potassium dihydrogen phosphate standards should be subjected to the same treatment as the unknown lipid sample. These inorganic phosphate standards are used to construct a graph of absorbance (based on reaction with ANSA) versus concentration.

The digested samples are treated with the ANSA reagent, and the mixture is placed in a boiling water bath until the blue color has stabilized at a maximum—usually 20 min. The absorbance of the samples (including the standards) are read at 830 nm. Then a working graph is prepared using the absorbance of the standards plotted against the phosphorus concentration of the standards. The concentration of the unknown lipid samples can be obtained from the working graph by using their absorbance values on one axis and correlating with the concentrations given on the other axis. A least-squares linear regression curve can be constructed and does aid in establishing the significance of the observations.

Thin-Layer Chromatography

A variety of precoated thin-layer chromatographic plates are available from commercial suppliers. The types described here are ones that have proven successful in the author's laboratory, but others could be equally adequate depending on the investigator's preference.

Pre-Washing

Silica gel G-precoated plates (250 μm; Analtech, New Jersey) are placed in a glass chromatography tank containing a filter paper lining (Whatman 1) and a solvent of chloroform–methanol–water, 65:35:7, v/v). The paper "wick" should be thoroughly wetted, prior to introduction of the plates. Then, the solvent is allowed to migrate to near the top of the plate. The plates are removed, air-dried in a fume hood, and placed in a desiccator until used. This prewashing removes organic material that might interfere with the detection of compounds in the lipid sample.

Sample Application

Prior to spotting of samples, the plate is scored lengthwise at 0.5- to 1.0-cm widths by use of a stainless steel needle. Excess silica gel is removed by gentle tapping of the plate, and the samples are ready for application. The outer lanes of the scored plate are reserved for standards (e.g., phosphatidylcholine, phosphatidylethanolamine, sphingomyelin, lysolecithin, and others as desired), which are readily available from several biochemical companies. Aliquots of the unknown lipid sample, which contain sufficient P, as detected by the qualitative spray above, are spotted on two to three lanes (usually in a volume ratio of 1:2:3) approximately 1.0 cm above the bottom of the plate. This is called the *origin*.

Chromatography Procedure

Then the plate is placed in the chromatography tank as described above. This tank contains a filter paper wick (saturated with solvent) and sufficent solvent in the tank to wet the plate approximately 0.5 cm above the bottom edge. The (top) lid is placed securely on the chromatography tank, and the solvent is allowed to ascend up the plate to within 1–2 cm from the top. In this solvent system and with these silica gel plates, the time for this operation is close to 50 min. The plates are then removed from the tank, and the solvent front is quickly marked on each lengthwise edge and placed in a fume hood to dry. These are the rudimentary steps to take for a single-dimensional thin-layer chromatography. In two-dimensional thin-layer chromatography, the procedure is different in that the lipid sample is placed as a concise spot at the lower left edge of the plate (approximately 1 cm from the bottom and 1 cm from the side). Then the plate is placed in a chromatography tank, usually with a neutral solvent, and allowed to run until the solvent is approximately 1 cm from the top edge. The plate is removed and air-dried in a fume hood. It is rotated 90° to the left and placed in a second chromatography tank, usually with an acidic solvent. The solvent is allowed to rise to near 1 cm from the top edge, and the plate is removed and air-dried in a fume hood. The two types of plates can be subjected to the spray reagents described below for location of compounds.

Detection Reagents

Now that a successful chromatogram has been run, the very important question is to evaluate in a qualititative sense the minimum number of compounds or classes of compounds present. The reason for this latter statement is that any visually single concise spot—for example, phosphatidylethanolamine—can be composed of at least three general species, namely, 1,2-diacyl, 1-O-alkenyl-2 acyl-, and 1-O-alkyl-2-acyl-. Yet all three species, together with several different types of long-chain hydrocarbon residues, migrate to the same R_f value. Hence a single concise spot is a beautiful sight indeed, but it can be a multifaceted beauty. The sprays described below are very useful and can be of great value if the results are not overinterpreted. This is just a simple reminder that these spray reagents give important clues but do not provide structure proof identification.

Through use of the following spray reagents, it is possible to obtain a reasonable assessment of the types of lipids present. These reagents have some degree of specificity and can provide a suitable reference point for further isolation and structural identification procedures. Only a limited number are discussed here.

Phosphorus and Char Reactions

These two tests can be particularly helpful. The P spray reagent was described above and can detect as little as 0.2 μg phospholipid P and yields a brilliant blue color usually within a minute. The char reaction is accomplished by spraying a separate plate (or the one after the phosphorus reaction) with concentrated sulfuric acid and then heating the plates on a hot plate at 180°C or in a muffle furnace at 400–600°C. The difference is of course that the char reaction (first a brown and then a definite charcoal color) is complete at the higher temperature within 5 min, whereas the reaction conducted at 180°C takes perhaps 30 min.

A comparison of the thin-layer chromatographic behavior of several lipid standards—for example, cholesterol (Ch), phosphatidylethanolamine (PE), phosphatidylcholine (PC), sphingomyelin (Sph), and lysophosphatidylcholine (LPC)—is presented in Figure 3-1.

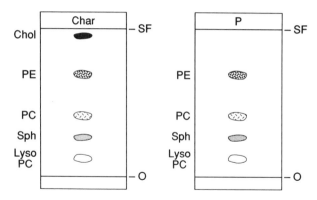

FIGURE 3-1. Thin-layer chromatographic profile of commonly encountered cellular lipids.

The solvent system for the run in Figure 3-1 was chloroform–methanol–water (65:35:7, v/v). The R_f values of these compounds can be calculated easily by measuring the migration distance of the sample from the origin and dividing this figure by the distance of travel of the solvent from the origin (O) to the solvent (SF). As a routine, the plates charred with sulfuric acid are photographed using a Polaroid MP-4 camera and the final print is attached to the laboratory notebook. The plates sprayed with the phosphorus stain do not photograph as well, and an alternative is to spray first with the phosphorus reagent; as the positive spots appear, these areas can be marked with a pencil. Then use of the sulfuric acid spray followed by charring can be used for photographic purposes.

Several other reagents, especially one called TNS, are very effective in the detection and partial identification of lipids. These are as follows.

TNS (6-p-Toluidine-2-naphthalene Sulfonic Acid)

This is a very effective reagent for the detection and quantitation of lipids on thin-layer chromatograms (Jones et al., 1982). The spray reagent is composed of 1 mM TNS in 50 mM Tris-HCl, pH 7.4. Subsequent to spraying, the plate is irradiated with ultraviolet light and within a very short period of time a fluorescent spot for each lipid (a bright spot on a dark background) will appear. TNS is nearly as sensitive as the phosphorus spray. It has the added advantage that it can be separated completely from the lipid by solvent extraction. This is accomplished by scraping the desired fluorescent spot into a tube and extracting with chloroform–methanol–water (1:2:0.8, v/v). Addition of chloroform and water to this extract, together with vortexing, will phase the TNS into the water-rich layer and the unaltered lipid into the chloroform-rich layer. This will be helpful in preparative thin-layer chromatography as will be described later.

Ninhydrin

This reagent is composed of 0.3% ninhydrin in 2-propanol–acetic acid–pyridine (90:10:4:10, v/v). The plate is sprayed at room temperature with this reagent, allowed to dry, and then placed in an oven at 100°C. Within 3–5 min at this temperature a purple color will indicate a positive reaction. Though this test is approximately two to three times less sensitive than the phosphorus spray, it is, nevertheless, an important adjunct in determining the presence of primary or secondary amines. However, a disadvantage is that the lipid sample cannot be recovered subsequent to the spray reaction.

Several other spray reagents used in detection of glycolipids and gangliosides are described in detail by Hakomori (1983).

Return to the Original Goal: Platelet Lipid Classes as Observed by TLC

An important (and early) objective of this chapter was to illustrate an approach to evaluation of the lipid classes present in a mammalian cell, such as the human platelet. Thus, armed with the information presented above, it is now feasible to start to explore this problem. As emphasized earlier, TLC can provide significant information on the types (or classes) of lipids present in a lipid extract analytical and/or structural technique to elucidate the chemical nature of the individual groups of compounds. A brief examination of the results that can be obtained in two types of thin-layer chromatography follows.

Single-Dimensional TLC

Using the total platelet lipid extract as described in the section entitled "Isolation of Human Platelets," single-dimensional thin-layer chromatography on a 250-μm-thick silica gel G-coated plate with a solvent system of chloroform–methanol–water (65:35:7, v/v) will yield the pattern shown in Figure 3-2.

The compounds represented in this facsimile were as follows: NI, neutral lipids (cholesterol, triacylglycerol); PE, phosphatidylethanolamine; PS, phosphatidylserine; PI, phosphatidylinositol; PC, phosphatidylcholine; Sph, sphingomyelin; X, gangliosides, polyphosphoinositides, lysophosphatidic acids. The chromatography was done on silica gel G plates with chloroform–methanol–ester (65:35:31, v/v) as the solvent.

As is evident from the patterns in Figure 3-2, the sulfuric acid (char) spray will detect all organic carbon-containing lipids whereas the phosphorus spray will detect only phospholipid phosphorus-containing lipids. In addition to these sprays, of course, the TNS reagent will react with all lipids and is very effective in locating various types of compounds on a plate.

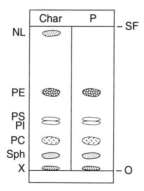

FIGURE 3-2. Separation of platelet lipids by thin-layer chromatography.

Another type of precoated thin-layer chromatographic plate that is particularly effective for a separation of lipids is Whatman K6. With this plate, a suitable solvent would be chloroform–acetone–methanol–acetic acid–water (4.5:2/1:1.3:0.5, v/v). However, a word of caution in the use of these plates is that it is not feasible to heat activate them prior to use due to a darkening effect under elevated temperatures. Hence, sulfuric acid charring is not recommended. Nonetheless, the TNS and phosphorus sprays are satisfactory detection agents.

In all of the experimental situations considered here, it is worthwhile to emphasize that none of these systems provide a sparkling clean separation of all the cellular lipids, especially in a single-dimensional run. For example, in the patterns shown in Figure 3-2, PS and PI are rather poorly separated. The areas where they are located on the plate can be removed by scraping, and this silica gel can be solvent treated and the compounds isolated as above and rechromatographed on plates impregnated with potassium oxalate. There will be a clean separation under these conditions. Of course an alternative is to subject the total lipid extract to chromatography on a similar-type (impregnated) plate, where PS and PI will be well separated. However, in the latter case, other lipids are less well separated, so again it comes down to what degree of separation is required whereby specific classes of phospholipids can be further separated and examined.

Polyphosphoinositides

The involvement of these phospholipids in signal transduction systems has been well documented. However, in the usual reaction, the amounts of compounds such as PIP_2 and PIP are very small (being less than 15% of the total inositol-containing phospholipids which may represent in the range of 3–8% of the total phospholipids). The most predominant phosphoinositide is phosphatidylinositol. Hence it is necessary to label cells with [^{3}H]inositol or $^{32}P_i$ to be able to detect whether these phospholipids are indeed involved in cellular signal transduction. This is accomplished by subjecting the total lipid extract to thin-layer chromatography on silica gel G-coated plates in a solvent system of chloroform–methanol–20% aqueous methylamine (60:36:10, v/v) (Shukla and Hanahan, 1983). On the basis of the migration pattern of standard PIP2, PIP, PI, and other phospholipids, it is possible to locate specific areas where the compounds of interest would migrate. These areas can be removed by scraping, extracted with chloroform–water. In the methylamine-containing solvent system the ascending order of migration (from the origin) is PIP2 → PIP → lysoPI → PI → PS + PA (not separated), → sphingomyelin → PC → PE.

Another approach to identification of the phosphoinositides by TLC was developed by Gonzalez-Sastro and Folch-Pi (1968). In order to avoid the possible influence of Ca^{2+} on the mobility of PIP_2, these investigators used silica gel H, which is devoid of $CaSO_4$ binder used in silica gel G plates, and included 1% potassium oxalate to bind any traces of Ca^{2+}. In a solvent system of chloroform–methanol–4N NH_4OH (9:7:2, v/v), the following R_f values were reported: PI, 0.78; PIP, 0.36; and PIP_2, 0.14.

Some years later, Billah and Lapetina (1982) employed silica gel G plates impregnated with 1% potassium oxalate containing 2 mM EDTA. In a solvent system of chloroform–methanol–4N NH_4OH (45:35:10, v/v), the following R_f values were observed: PIP_2, 0.28; PIP, 0.38; lysoPI, 0.44; and PA, 0.51. The other commonly encountered phospholipids such as PC, PE, and Sph would migrate at still higher R_f values.

Phosphatidic Acid and Lysophosphatidic Acid

It was established several years ago that phospholipids such as phosphatidic acid and lysophosphatidic acid exhibited biological activity. Recent excellent evidence has accumulated supporting their role as potent agonists in cellular signaling. Thus, it is important to establish the presence of these compounds in a lipid extract, and this can be accomplished by thin-layer chromatography. Using a high-performance silica gel 60-coated plate with a solvent system of a benzene–pyridine–formic acid (50:40:11, v/v), Tokumura et al. (1986) observed the following R_f values for several phospholipids:

PA	0.63	Sph	0.10
PE	0.47	lysoPC	
lysoPA	0.40	lysoPI	0.00 (origin)
PS	0.32	PIP_2, PIP	
PC	0.22	PI	

Two-Dimensional Thin-Layer Chromatography

The use of two-dimensional thin-layer chromatography for separation of platelet lipids will provide the profile exhibited in Figure 3-3.

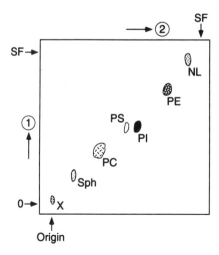

FIGURE 3-3. Two-dimensional thin-layer chromatography of platelet lipids.

The chromatograph in Figure 3-3 was obtained by applying a platelet lipid sample to the lower left section of the plate (approximately 1 cm from the bottom and 1 cm from the side). The initial solvent, chloroform–methanol–water (65:35:7, v/v) was run in the direction of the arrow (1). After the solvent had ascended to within 1 cm of the top, the plate was withdrawn from the chromatography tank and air-dried in a fume hood. Then the plate was turned 90° and placed in a tank containing chloroform–methanol–4N NH_4OH (65:34:4 v/v), and the solvent was allowed to ascend in the direction of the arrow (2). When the solvent had reached close to 1 cm from the top, the plate was withdrawn and allowed to air dry in a fume hood. It then can be sprayed with a reagent such as the TNS solution and viewed under ultraviolet light. The symbols for the individual groups of compounds are the same as described in the section entitled "Single-Dimensional TLC."

A Brief Recap

It should be quite clear by now that separation and identification of lipids in general, and phospholipids in specific, can at once appear disarmingly simple. Yet proof of structure and an ultimate well-defined analysis of lipids in a cell or a reaction system can be grudgingly difficult. Perhaps the most important point centers on the goal(s) of the research project and how much detailed information is needed.

In the usual instance it is necessary to use more than one analytical tool for identification (and separation) of the phospholipids in a biological extract. Other chromatographic techniques of value will be discussed later. However, now it is worthwhile to describe a methodology by which a nearly quantitative assessment of the lipid P distribution in a sample can be determined on a thin-layer chromatogram.

Quantitative Phospholipid Phosphorus Assay of Thin-Layer Chromatograms

It is possible to conduct a nearly quantitative analysis of the phosphorus distributed in the various classes of phospholipids separated on a thin-layer chromatogram. Usually two-dimensional chromatography is performed, though a similar approach can be accomplished in a single dimension. The basic approach is to lightly spray the plate with the phosphorus detection reagent described in the section entitled "Qualitative Test for Lipid P." The positive spots are removed by scraping into a digestion tube. A comparable adjacent area of non-phosphorus-positive absorbent is removed and used as a background control. Actually prior to spraying the plate, an aliquot of the total platelet lipid should be applied to the lower right section of the plate. Subsequent to the phosphorus spray, this spot is removed by scraping and serves as a control for calculation of recovery. Then the individual samples are digested as described earlier with perchloric acid, and the color is developed as noted. The samples are centrifuged to sediment the silica gel, and the absorbance is read as usual. Recovery is usually in the range of 90–95%.

Other Separation Techniques of Value

Preparative Thin-Layer Chromatography

Whereas the usual so-called analytical plate (e.g., 250-μm-thick silica gel G or H) can accommodate 1–2 mg total lipid (without serious smearing due to overloading), it is possible to obtain (from commercial houses) plates coated to 1000-μm thickness. As a result, upwards of 30–40 mg total lipid can be separated on a single plate measuring 20 cm × 20 cm. Usually single-dimensional chromatography is used. A serious drawback to the use of these very thick plates is the "sagging phenomena." Even though the lipid sample may be applied very evenly and neatly across the origin, the bands often will move uneveningly or "sag." This usually occurs in the middle of the plate and may simply mean that the plate is not uniform in thickness.

A better separation can be effected using a 10-cm × 20-cm plate of 1000-μm thickness (if available). In general, however, one can anticipate better results if a thinner plate, say 500 μm thick, is employed. So even if it is necessary to use four or five plates, the results are usually much, much better. A TNS spray reagent can be used to identify the desired bands; these areas are then removed by scraping and are subjected to extraction using the Bligh-Dyer solvent system of chloroform–methanol–water (1:2:0.8, v/v). As discussed earlier in this chapter, if any acidic compounds, such as lysophos-phatidic acids, are present in the lipid sample and if the chromatography is conducted on a silica gel G plate, the solvent should contain acid.

Preparative thin-layer chromatography, therefore, has much merit as a separation technique and should be considered seriously as an excellent tool in isolation of sufficient quatities of particular phospholipids for further chemical or biochemical study.

Silicic Acid Column Chromatography

In the 1960s and 1970s, it was *de rigeur* to use silicic acid column chroma-tography to separate lipid mixtures into their component parts. Certainly at that stage of development of the field of modern lipid chemistry and bio-chemistry, it was mandatory to obtain milligram quantities of individual classes of lipids for structure proof studies. It was a potent mehodology due to the commercial availability of high-quality and uniform-size silicic acid. This approach aided immeasurably in establishing the chemical structure of many cellular phospholipids. It would be remiss not to mention other absorbents in use at that time for separation of certain lipids—for example, Florisil, Se-phadex, Factice, and ion exchangers. However, by and large, silicic acid column chromatography dominated the field and was a highly successful technique. Nonetheless, as the field progressed and the structural characteris-tics of most of the cellular phospholipids were defined, the need for this type of chromatography was diminished. This change in direction was stimulated by the increased use of thin-layer chromatography and high-performance liquid chromatography.

Even though the column chromatographic technique is not as popular as it was several years ago, it still has some very useful applications. Only two examples will be cited here:

Chromatographic Bulk Separation of Neutral Lipids from Phospholipids and Specific Fractionation of Components

BULK SEPARATION.   This can be accomplished in a relatively few steps as follows:

A glass column fitted with a glass/Teflon stopcock and containing a glass wool plug at the bottom of the column is filled with a slurry of Silicar CC-7 or SilicAR CC-4 (Mallinckrodt) in *n*-hexane (or petroleum ether, b.p. 30–60°C)–diethyl ether (1:10, v/v). The top of the glass column is usually fitted with a ground glass ball and socket, with the top piece containing a glass reservoir for solvent. The top of the solvent reservoir has a ball-and-socket joint through which nitrogen under pressure can be introduced. Normally nitrogen pressure is used to pack the column to a constant height and reduce the head of solvent on adsorbent to a very small volume so that the sample can be evenly introduced. A height-to-diameter ratio of 10–15 is desirable.

The sample, containing 1 mg of lipid P per 2 g of silicic acid, is carefully applied and then additional solvent of hexane (petroleum ether)–diethyl ether (1:10, v/v) is introduced and elution started under pressure. The progress of the elution of the neutral lipids, which will come through in this solvent, can be monitored by placing an aliquot at intervals on a mini-silica gel plate and spraying the spot with sulfuric acid. Subsequent heating in a muffle furnace or on a hot plate will give visual evidence of organic material. Usually 15 column volumes of solvent will elute all the neutral lipid.

After elution of the neutral lipids, the phospholipids can be recovered by using a solvent of chloroform–methanol (1:10, v/v). Progress of the elution can be monitored by using the phosphorus spray combined (if desired) with the sulfuric acid char reaction. Approximately 15–20 column volumes will remove all the phospholipids.

The value of this approach is that neutral lipid and phospholipid fractions can be separated very quickly and then examined cleanly separated from each other. Columns as small as 1 g and as large as 200–1000 g have been employed in this type of separation procedure.

SPECIFIC FRACTIONATION.   This can be accomplished for the neutral lipid fraction and the phospholipid by use of separate silicic acid columns. The usual profile for these two chromatographic runs would be as follows:

Neutral Lipids.   The neutral lipid fraction, free of phospholipids, will give the following elution sequence with the indicated solvents:

Solvents	Component(s) eluted
n-Hexane	Hydrocarbons
n-Hexane–benzene (8.5:1.5, v/v)	Sterol esters
n-Hexane–diethyl ether (95:5, v/v)	Triacylglycerols, free fatty acids
n-Hexane–diethyl ether (85:15, v/v)	Free sterol
n-Hexane–diethyl ether (70:30, v/v)	Diglyceride
n-Hexane–diethyl ether (10:90, v/v)	Monoglycerides

Progress of the elution can be monitored by placing aliquots of the eluent on a mini-silica gel G plate, spraying with sulfuric acid and then charring. This is only one example of a solvent system that works for this particular situation. Many other combinations could be tried; and again, as stated many times, it depends on the goal of the particular experiment.

Phospholipids.   This fraction free of neutral lipids can be applied to a SilicAR CC-7 or CC-4 column, with a loading of 0.5 mg of phospholipid P per gram of absorbent and a height-to-diameter ratio of 15–20. Again the solvent systems to be used in this separation will depend on the objective of the experiment. As would be expected, there are many solvent systems that have been used by many different investigators, and each one has merit on its own. One that has been particularly successful in the author's laboratory is as follows:

Solvents	Component(s) eluted
Acetone–methanol (9:1, v/v)	Glycolipids, gangliosides, phosphatidylinositol
Chloroform–methanol (4:1, v/v)	Phosphatidylethanolamine
Chloroform–methanol (2:1, v/v)	Phosphatidylserine
Chloroform–methanol (7:6, v/v)	Phosphatidylcholine, sphingomyelin
Chloroform–methanol (1:10, v/v)	Lysolecithin

This is a respectable type of separation, and yet it is obvious that not all compounds or classes of compounds are separated precisely and neatly from each. There simply is not such a magic system, and one must realize that further fractionation may be necessary. It is interesting to note that phosphatidylcholine and sphingomyelin tend to blend together. Usually the more unsaturated phosphatidylcholine species elute early on, followed by the more saturated one. By judicious handling of solvent eluate volume (by a fraction collector), a quite reasonable separation can be obtained.

Aluminum Oxide Chromatography

Though now approaching the venerable age of 46 years, aluminum oxide as a separation media deserves consideration (Hanahan et al., 1951). It is particularly effective in separating the choline-containing from the non-choline-containing phospholipids. However, prior to such a chromatographic intervention, it is strongly recommended that any total lipid extract containing the above phospholipids be passed through a silicic acid column to remove the neutral lipids. Otherwise the neutral lipids will be eluted together with the choline-containing phospholipids on aluminum oxide column chromatography.

When the neutral lipids have been removed, the recovered phospholipids are dissolved in chloroform–methanol (1:1, v/v) and then placed on an aluminum oxide column (neutral or slightly basic, activity grade 1), at a loading ratio of 1 mg of phospholipid P per gram of aluminum oxide. The column height-to-diameter ratio is approximately 10, and a glass column fitted with a Teflon glass stopcock is recommended. Then the following "elution" program is followed:

Solvent I (chloroform–methanol 1:1, v/v). In 10 column volumes, all of the choline-containing phospholipids should have passed through. Then, start the following:

Solvent II (ethanol–chloroform–water, 5:2:2, v/v). This will elute the remaining non-choline-containing phospholipids again in approximately 10 column volumes of solvent.

The progress of elution is monitored by removing 50- to 75-μl aliquots of the eluate, spotting on a mini-silica gel (250 μm) plate, and then spraying with the phospholipid P reagent and/or sulfuric acid (with char reaction), as described earlier in this chapter.

The advantage of this type of chromatography is that one can do, subsequently, a more refined chromatographic separation on a sample with fewer components and in a faster time period. Certainly a reasonable chromatographic separation of a total lipid extract can be achieved on a single column of silicic acid, but it has potentially frustrating aspects in that fractions will tend to overlap more, there is more likelihood of degradation of sensitive phospholipids (such as the vinyl ether type) with longer time periods on the adsorbent, and there is marked slowing in the elution rate as one changes to more polar solvents.

The choice of the type of aluminum oxide is important, and it is suggested that it have a pH near 7.0–8.0 with an activity grade of 1. If too acidic or basic, this adsorbent can have deleterious effects on certain phospholipids. Aluminum oxide (alumina) suitable for chromatography is available from commercial biochemical supply houses.

Finally, two points of interest need to be discussed. First, this is not a typical adsorption/elution-type chromatography since due to their zwitterionic characteristic, over the entire pH range, the choline-containing phospholipids pass rapidly through the aluminum oxide with little or no detectable adsorption. On the other hand, the non-choline-containing phospholipids, such as PE, PS, PI, PIP, PIP_2, PA, and lysoPA, with a distinct charge, are adsorbed in Solvent I. Only with the application of another solvent mixture, such as ethanol–chloroform–water (5:2:2, v/v), can they be eluted from the adsorbent. In actual fact, the successive use of ethanol–chloroform–water (5:2:1, v/v) and then ethanol–chloroform–water (5:2:2, v/v) will allow a reasonable separation of PE from the other phospholipids. Just be aware of the fact that the elution rate with these last two solvents, even with nitrogen pressure, is very, very slow. A second point of interest centers on the possibility that the choline-containing fraction can contain other analogous compounds such as phosphatidyldimethylethanolamine and the phosphate derivatives. Though the latter, at least in mammalian tissues, are usually found in very small amounts, forewarned is forearmed.

High-Performance Liquid Chromatography (HPLC)

This novel and sophisticated technique has become very useful as a tool in the separation and identification of lipids. Both neutral lipids, especially the triacylglycerols and the phospholipids, can be separated by this procedure.

Two brief, but very definitive, reviews on the HPLC of triacylglycerols (Plattner, 1981) and of phospholipids (Porter and Weenan, 1981) are recommended reading.

High-performance liquid chromatography, also called high-pressure liquid chromatography, represents a liquid chromatographic technique utilizing high inlet pressure and high sensitivity. In the silicic acid column chromatographic approach described earlier, vertical columns are used, usually on an open bed (or if closed, at very low applied pressures). In HPLC, the stationary phase, at least for phospholipids and triacylglycerols (triglycerides), is a solid; a solution (solvent) is pumped through the column at very high inlet pressures, sometimes in the range of 3000–4000 psi. Usually, the pressures are much lower near 400–800 psi. Thus the phospholipid mixture is adsorbed onto a special column, and subsequently each class of phospholipid (PE, PS, PI, PC, etc.) is displaced at different rates by a single solvent system.

As regards the lipids, two types of adsorbents are available, one of which is a form of silica gel and is utilized in normal-phase HPLC, and the other of which can be a silica gel bonded to a hydrophobic chain and is employed in reverse-phase HPLC. In normal-phase HPLC the phospholipids appear to be separated based on the molecular classes present (PE, PC, Sph, etc.), whereas in reverse-phase HPLC the separation is closely related to the lipophilic character of the acyl (fatty acyl, hydrocarbon chain) of the particular phospholipids. High-quality adsorbents suitable for HPLC are easily available from commercial companies.

There are excellent HPLC systems available on the market today, yet there is one area of concern with this instrumentation, and this rests with the detection units. Certainly the most widely used detector system employs a low dead-volume micro-ultraviolet detector. This latter unit operates near 200 nm and detects mainly unsaturated linkages in phospholipids (or lipid) samples. Some contribution by carbonyl functions can be expected. This approach is an advantage when the sample under study contains olefinic groups, but will not detect those with saturated side (hydrocarbon) chains. An alternative detector is the refractive index monitor which is often called a universal detector, since it is based on the concept that the refractive index of the solvent changes when a solute is present. The drawback of the latter unit lies in its sensitivity, which is approximately 15- to 20-fold less than that of the ultraviolet monitor.

A typical separation of platelet phospholipids can be achieved as follows: A sample containing 40–50 μg of phospholipid (by weight) is applied to a μ Porasil (polar-type) column (Waters, Milford, CT) measuring 7×30 cm, and a solvent system of acetonitrile–methanol–45% phosphoric acid (130:5:1.5, v/v) is applied, as suggested by Chen and Kou (1982). The flow rate was 1 ml/min, and the elution was monitored by an ultraviolet detector sensitive at 203 nm. The total elution time in this case was 40 min, and the profile is illustrated in Figure 3-4.

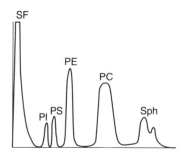

FIGURE 3-4. High-performance liquid chromatographic separation of platelet phospholipids.

The individual fractions can be collected and subjected to further examination, for example, by fast atom bombardment (see the next section on mass spectrometry) or assayed for biological activity, especially if platelet activating factor is present in the original sample. Interestingly, in the latter instance PAF would not be detected by the ultraviolet detector since it usually contains very low amounts of olefinic double bonds.

Nonetheless, it has been shown to elute immediately after the PC peak. LysoPC, if present, would elute between the PAF and sphingomyelin (Sph) peak. The latter component often shows a double peak, and this can be attributed to the separation of distinct fatty acyl species. It is well to emphasize again that compounds such as PAF and lysoPC have very low levels of unsaturated bonds present, and hence a detector other than an ultraviolet monitor would have to be used. An alternate approach would be to use tritiated PAF or tritiated lysoPC as examples and assay the eluates by liquid scintillation counting. Since the labeled compounds are at tracer dose levels, one could still assay for biological activity associated with the compounds.

A Potent Structure Proof Technique: Mass Spectrometry (Fast Atom Bombardment)

Mass spectrometry, especially in the form of fast atom bombardment, has become a potent tool in the elucidation of the structure of phospholipids. In this technique, the phospholipid sample in amounts in the range of 15–100 ng (or more), dissolved in chloroform–methanol (1:1, v/v), is mixed with a glycerol or thioglycerol matrix and placed on a sample mount in the spectrometer. A fast atom gun is then aimed, at a 90° angle, at the sample and the resulting ionized particles are directed to a mass analyzer unit. Under these conditions, most of the energy of the beam is limited to the surface of sample matrix. Consequently, the bulk of the sample is unaltered and can be recovered by solvent extraction. In the case of phospholipids, a mass ion, MH^+, is produced; also, some fragments are useful for structural analysis. A more detailed description of this particular methodology can be found in a book by Watson (1985) and in a review by Hanahan and Weintraub (1985). The latter article centers specific attention on the structural proof analysis of platelet activating factor (1-*O*-alkyl-2-acetyl-*sn*-glycero-3-phosphocholine) utilizing organic chemical procedures together with fast atom bombardment and other mass spectrometric techniques.

In a typical examination of a phospholipid sample using the fast atom bombardment technique, the following experimental protocol is practiced in our mass spectrometry laboratory under the direction of Dr. Susan T. Weintraub. Spectra are acquired on a Finnegan-MAT Model 212 mass spectrometer in combination with an INCOS data system. An Ion-Tek saddle field atom gun was employed with xenon gas at a voltage of 9 kV. The ion temperature is maintained at 60°C and the accelerating voltage is 3 kV. A sample containing 1 μg of phospholipid is applied to the copper FAB probe tip in 2–3 μl of chloroform–methanol (2:1, v/v); approximately 2 μl of thioglycerol is added and mixed with the phospholipid sample. The contribution of the thioglycerol matrix is subtracted from each spectrum prior to evaluation of the phospholipid sample spectrum. A typical spectra is described as follows, for 16:0 PAF (1-*O*-hexadecyl-2-acetyl-*sn*-glycero-3-phosphocholine):

In a typical spectrum, the peak at 524 represents the mass ion MH$^+$ and m/z 482 represents the deacetylated derivative. Also an additional fragment of importance in use in structure studies is the m/z 184, which represents the *O*-phosphocholine residue. These are the major fragments obtained from 1-*O*-hexadecyl-2-acetyl-*sn*-glycero-3-phosphocholine. Similar information can be derived from a similar derivative in which the acetyl group is replaced by an oleoyl residue. In this case the mass ion MH$^+$ would be 746 and m/z 184 would represent the *O*-phosphocholine group. If a linoleoyl (at *sn*-2), hexadecyl (at *sn*-1)-substituted derivative were present, then a mass ion of 744 would be detected. With regard to the polar head groups, an ethanolamine phospholipid would yield on FAB an m/z 142, indicative of a protonated *O*-phosphoethanolamine.

The question can be posed as to whether the stereochemical conformation of a phosphoglyceride can be determined by FAB-MS. While no rigorous examination of synthetic *sn*-3 or *sn*-1 compounds by this technique (or any other mass spectrometric procedure) has been reported, it seems highly unlikely that any detectable or significant differences would be noted. Nonetheless, FAB-MS and other forms of mass spectrometry are powerful tools useful in the structural identification of organic compounds, and this possibility may develop.

A Transition

The goals of the first three chapters have been to prepare a basic foundation for the reader on the general chemical characteristics of phospholipids and to outline some approaches to their isolation and detection. The overture is completed and now it is time to proceed into the main movement, a more in-depth discussion of the specific chemical nature of several classes of cellular membrane phospholipids. It is hoped that the information detailed in the succeeding chapters will be of value to an investigator studying the physiological or biochemical behavior of these exciting compounds. While the underlying theme will be to relate the information provided to signal transduction systems, it is obvious that it will be applicable as well to many other experimental protocols.

FOUR

CHOLINE-CONTAINING PHOSPHOLIPIDS

Diacyl-, Alkylacyl-, and Alkenylacylcholine Phosphoglycerides and Sphingomyelin

The Reason

In choosing the order for discussion of phospholipids, it is not the intention to single out one particular group as the most important; rather, an initial premise would be that all phospholipids are critical to a cell's structure and metabolism. Certainly, as has been emphasized before, phospholipids have been shown to have key roles in the process of cellular signal transduction, and it is debatable which of several types of phospholipids is the most important. There is no doubt that the mechanism of involvement of membrane phospholipids in these complex reactions has presented a major experimental challenge, and as such this has titillated the acute scientific senses of many researchers. It is equally true also that an important field of study is emerging in cell signaling, in which unusual cellular disorders have been noted. Certainly the latter will implicate alterations or aberrations in membrane phospholipid chemistry and metabolism in one way or another. This digression was made to show quite simply that it behooves one to understand the chemical/biochemical characteristics of the phospholipids in order to best meet the challenges of this field (and, of course, other related ones as well).

On the basis of undoubted faulty logic on the choice of the order of topics, one simply can retreat to the argument of personal preference. Thus, the first group of compounds will be the choline-containing phospholipids—that is, the choline phosphoglycerides and the choline sphingolipids. As it so happens, these are among the most ubiquitous phospholipids in nature and, at least in the early "chemical" years of investigations on the phospholipids, the best-studied group.

Sequel to a Premise

It is assumed at this junction that a highly purified phospholipid has been obtained, usually through the use of chromatographic procedures. A frequently asked question is, How do I tell whether the sample is pure? It is a logical question, especially with compounds isolated from naturally occurring sources. In actual fact, there is no simple answer. Perhaps the best way is to use as many criteria as possible to establish the character of the material under study. A few suggestions are as follows:

Thin-Layer Chromatographic Behavior

The objective is to evaluate the migration behavior of the sample using three different solvent systems. One of these would be a neutral type, namely, chloroform–methanol–water (65:35:7, v/v). A second would be an acidic solvent, such as chloroform–acetone–methanol–acetic acid–water (4.5:2:1:1.3:0.5, v/v). The third one would be a basic solvent system, using chloroform–methanol–28% ammonia (80:20:2, v/v). Then, using different spray systems, such as for phospholipid P, for organic material (char reaction), and for free amino groups (ninhydrin), a very real sense of the homogeneity of the sample can be obtained through this approach. In the case of a neutral compound such as phosphatidylcholine (zwitterionic over the entire pH range), there will be only a limited difference in R_f values in these three solvent systems. On the other hand, a compound such as phosphatidyl ethanolamine, which definitely has a titratable (polar head) group, will show a dramatic shift in its R_f value in two of these solvents. For example, in the neutral solvent, phosphatidylethanolamine might show an R_f value [using a silica gel G (250 µm) plate) near 0.60. If its R_f is checked in a basic solvent system, the value could be near 0.10. These maneuvers together with the use of several different group specific sprays will provide a reasonable comfort level to an investigator with regard to the purity of the sample under study.

FAB-MS Spectral Pattern

Even though this is a reasonably sophisticated technique (and one must weigh this point with the cost of instrumentation, maintenance and the need for a highly skilled mass spectrometrist), it nevertheless is a very potent tool in assessing the structural nature of a phospholipid. In Chapter 3 a typical spectrum of a phospholipid was described; the information gathered from this pattern can be of enormous help in deciphering the structure.

However, as impressive as this methodology can be it is important to stress that the technique will not provide an absolute guarantee of purity. Certainly a 5% level of another phospholipid most likely would not be detected. As a consequence, one should be aware of this possibility. Also it cannot be regarded as a quantitative procedure due to the nonquantitative behavior of the phospholipid with a matrix, such as thioglycerol. If further general information is needed on this subject, reference can be made to the review by Hanahan and Weintraub (1985).

A Few General Analytical Techniques

If one wishes to explore further the chemical nature of the phospholipid sample, there are a few relatively simple methods that can be used. Usually the phospholipid is obtained as a viscous material (either colorless or a light tan color) but seldom, if ever, as a friable powder. As a consequence, the most effective way to handle the sample is to dissolve it in a solvent such as chloroform–methanol (2:1, v/v) and make to volume in a glass-stoppered volumetric flask. Then aliquots can be taken for the following simple analytical procedures:

Total Weight Determination

This is accomplished by using carefully tared, 5- or 10-ml Erlenmeyer flasks, into one of which is transferred a 50- to 100-μl aliquot containing approximately 5–10 mg. Using clean clamplike tongs, these flasks are placed in a vacuum oven operating at 30°C. A vacuum (from a house line) is applied slowly and the flasks are left *in vacuo* for 3–4 hr. Carefully release the vacuum, remove the flasks, and weigh against each other. The difference in weight obviously will represent the amount of phospholipid in the aliquot. Normally, triplicate samples are run if an accurate result is desired.

Of course this method requires much larger amounts of lipid than ever found, for example, in an extract from a signal transduction experiment. In the latter instance, the most expedient approach would be to use a total phosphorus value as the basis for calculation of the probable weight of the particular fraction. This will be explained in the next section.

Total Phosphorus Assay

Aliquots containing 1–5 μg of phosphorus are transferred to digestions tubes, the solvent is evaporated in a fume hood, and digestion and assay are conducted as described in Chapter 3. Then the total phosphorus content of the sample coupled with the total weight value can give you a percent P value.

However, if there is an insufficient amount of material available to run a total weight determination as described above, then a qualitative estimate can be obtained by multiplying the P value by 25. The latter number is derived by dividing the molecular weight of a known phosphoglyceride (e.g., palmitoyl oleoyl glycerophosphocholine, molecular weight 754) by the molecular weight of phosphorus, 31. Hence a total of 20 μg of phosphorus would give 500 μg total weight. It should be noted that this is a rather ancient way to calculate the probable weight of a phospholipid sample; as such it must be regarded as a "guesstimate," but it does provide information of some value.

Total Nitrogen Assay

In this procedure the lipid sample, containing in the range of 2–15 μg of nitrogen, is placed in a digestion tube and the solvent is removed in a fume hood. The residue is then digested in perchloric acid and treated with a phenol–nitroprusside mixture plus a sodium hypochlorite solution. A blue color is produced and the absorbance of the solution can be read at 635 nm. Further details of this assay can be obtained in an article by Sloane-Stanley (1967). With a nitrogen value (as well as phosphorus value) in hand, a calculation can be made to give an N/P molar ratio. In the case of di-acylphosphatidylcholine, this number would be 1.0.

This assay is really meant to be used in experiments in which milligram quantities of lipid are available for an initial structure proof of an unknown compound. Obviously, it is not employed in lipid preparations from small quantities of cells (e.g., 2.0×10^8 cells/ml), used in signal transduction experiments. There simply is not sufficient sample for such an analysis. An alternative approach, if enough material is available, is to have a C, H, N, P analysis performed by a commercial organic analytical laboratory. As mentioned, the major difficulty is that the lipid is often a very viscous material, and this makes reproducible sampling a problem. Also, in the author's experience, the techniques used by commercial laboratories give lower-than-expected results even with synthetic saturated phospholipids. It may have to do with the combustion techniques used or unusual behavior of these compounds under assay conditions. The exact reason for this behavior is not clear.

Fatty Acid Ester Determination

A convenient route to measurement of the ester bond content of a phospholipid is through cleavage of the ester by alkaline hydroxamic acid to yield a hydroxamate. Upon mixing this derivative with ferric perchlorate, a color is formed and the absorbance of the resulting solution can be read at 530 nm. The method can measure from 0.2 mg to 2–3 mg of ester. Further details on the methodology can be obtained by reference to a paper by Snyder and Stephens (1959). Again, this assay is most useful for investigations wherein milligram quantities of lipid are available.

One can calculate quite easily the ester-to-P molar ratio, to give a fatty acid ester-to-P value. For a diacylphosphatidylcholine, this number would be 2.0, whereas for a monoether, monoacyl phospholipid the value would be 1.0.

Another technique for determination of the fatty acid ester content of phosphatidylcholine is through the use of infrared spectrometry. The fatty acid ester groups in the native phospholipid and the methyl esters derived from this phospholipid absorb quite strongly at 1750–1730 cm^{-1}. This latter absorption is intimately associated with the carbonyl function. In a concentration range of 10–100 μg/ml solvent, a Beer-Lambert relationship is noted and a straight-line relationship between concentration and absorbance is achieved. Thus, using a pure diacylphosphatidylcholine and readily available pure methyl esters, a standard curve (i.e., concentration versus absorbance) can be constructed and the ester content of an unknown preparation can be determined. The concentration range noted above can be extended to a significant degree through use of Fourier transform infrared instrumentation.

The above methodology can provide a considerable amount of information on the chemical nature of the sample under study. However, again, it will not prove completely the structure of the phospholipid in question. With this point in mind, it is appropriate to examine in detail the chemistry of one of the most prevalent phospholipids in mammalian cells, namely, diacylglycerophospho-choline (phosphatidylcholine). A primary goal is to provide some insight into the manner by which its structure can be proven and how this information can be applied to investigations concerned with phospholipid behavior in membranes. Although attention will be focused first on the classic (diacyl) phosphatidylcholine, the closely related analogs, the vinyl ether (plasmalogen) form and the saturated ether type will be discussed in some detail.

The Ubiquitous One: (Diacyl)Phosphatidylcholine

This phospholipid is a most prominent member of the phosphoglyceride class of compounds, which frequently account for 50–60% of the total lipid in a cell (or a membrane).

Its structural formula can be represented as shown in Figure 4-1, where R_1 and R_2 represent long chain (13–21) hydrocarbon residues. Certain important features of this structure should be emphasized at this point:

Optical Asymmetry

This derives from the asymmetric carbon atom at position C-2. On the basis of the nomenclature adopted in Chapter 1, this compound has an sn-3 configuration. In all the diacylphosphatidylcholine molecules found in mammalian cells and tissues, this is the only stereochemical configuration present. No evidence for the sn-2 or sn-1 configuration has been reported.

Positional Asymmetry

A very consistent chracteristic is the selective positioning of the fatty acyl residues on the glycerol backbone. Thus, in a very large percentage of the naturally ocurring phosphatidylcholines, the hydrocarbon chains on the C-1 position are highly saturated—hence little or no olefinic bonds are present. On the other hand, when the acyl residues on C-2 are analyzed, well over 95%

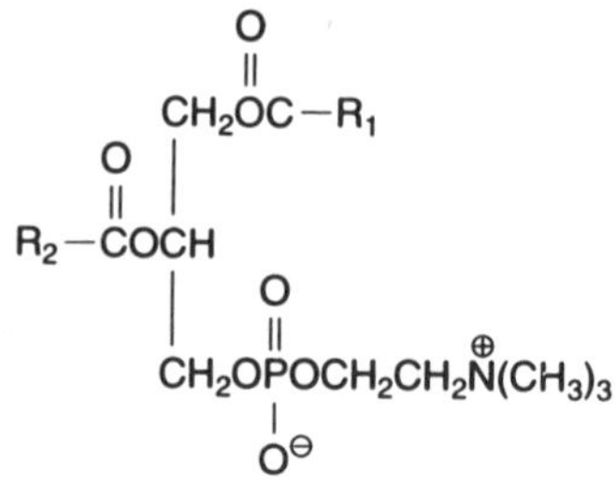

$$
\begin{array}{c}
\overset{\displaystyle O}{\overset{\displaystyle \|}{CH_2OC-R_1}} \\
\overset{\displaystyle O}{\overset{\displaystyle \|}{R_2-COCH}} \\
\overset{\displaystyle O}{\overset{\displaystyle \|}{CH_2OPOCH_2CH_2\overset{\oplus}{N}(CH_3)_3}} \\
\overset{\displaystyle |}{O^{\ominus}}
\end{array}
$$

FIGURE 4-1. Phosphatidylcholine, a commonly encountered cellular phosphoglyceride.

are represented by unsaturated (olefinic) chains. Of course, as expected, there are always exceptions to this grand rule; for example, dipalmitoyl phosphatidylcholine is the main species in lung surfactant. Nonetheless, in general, the positional asymmetry of the fatty acyl groups, with saturated types in the C-1 position and unsaturated types in the C-2 position, is maintained in most phosphatidylcholines examined to date.

Glycerol Backbone

This alcohol is the main constituent of the mammalian-derived phosphatidycholines. The only exception has been the report of a diol present in very small amounts (as a phosphocholine derivative) in lung tissue.

Polar Head Group

Of course as the name of this phosphoglyceride implies, a phosphocholine group is present. Interestingly the nitrogen base, choline, as represented in Figure 4-2, has a pK near 13.9 and forms an ester bond to the phosphate group. The latter is then bound in ester form to the glycerol. This association produces a polar head group with a strong zwitterionic character. On the basis of titrimetric assays, phosphatidylcholine is clearly considered as a zwitterion over the entire pH range.

An Approach to Structure Proof on Phosphatidycholine

The proof of structure on the naturally occurring phosphoglycerides is by no means a simple task, but it can be done with perseverance. Perhaps the best way to initiate such experiments is to explore in order the topics briefly discussed in the section entitled "The Ubiquitous One: (Diacyl)Phosphatidylcholine." These are the essential components of a structure proof for these compounds.

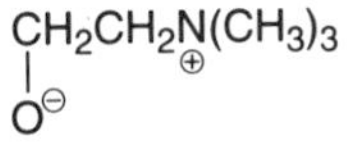

$$\underset{\overset{|}{\overset{|}{O^{\ominus}}}}{CH_2CH_2\overset{\oplus}{N}(CH_3)_3}$$

Choline

FIGURE 4-2. Choline, a constituent of the polar head group of phosphatidylcholine.

One may ask why the FAB-MS technique, described in Chapter 3, would not save the day by simply obtaining the mass spectral pattern. As noted before and reemphasized here, FAB-MS is a powerful method and will give important information relative to the structure, but will not give a complete structure proof. First, it will not provide information on the stereochemical form, nor will it provide data on the positioning of the fatty acids on the presumed glycerol backbone and will not indicate the purity of the sample. Nonetheless, it is a masterful technique and if one has access to such sophisticated equipment, use it.

The following discussion will assume that sufficient sample is available on which to undertake a complete structure proof. However, in the case of an investigation utilizing membranes, it obviously will be necessary to modify the experimental design to accommodate to the size of the sample. A description of suggested experimental approaches will be discussed; and while the importance of certain procedures will be stressed, it will be up to the investigator to decide on the *modus operandi* based on sample size.

It is assumed at this point that sufficient material and data have been accumulated, via techniques such as thin-layer chromatography, P analysis, FAE/P values, and an FAB-MS spectrum (if available), to suggest that the phospholipid is a diacylphosphatidylcholine or a closely related derivative.

Stereochemical Form: General Comments

The presence of an asymmetric center in the sample can be derived from its behavior in solution to plane-polarized light. This can be accomplished through use of an automatic polarimeter (Perkin-Elmer Model 241, sodium lamp). It is strongly suggested that such equipment, rather than visual examination, be used to measure the optical activity of the phospholipids, which are notoriously "low rotators." Phospholipids such as sn-3-phosphatidylcholine have a specific rotation, $[\alpha]_D^{26}$, value of $+6°$ to $+7°$. Comparison with the $[\alpha]_D^{26}$ value of glucose, $+47.9$, immediately defines the problem. Prior to the availability of an automatic polarimeter, the visual approach was really an act of faith. As expected, concentrated solutions were required; and if they had a slight tan color, then differentiation between the two half-fields in the polarimeter was tenuous at best. So, if optical activity values are desired, it behooves one to use an automatic polarimeter.

So the question is how to proceed to establish the stereochemical structure of a naturally occurring phosphatidylcholine. Two approaches lend themselves to such an analysis. The first is a direct chemical attack, and the second is a biochemical (enzymatic) approach. The primary goal is to prepare non-racemized fragments, such as glycerophosphocholine, a diglyceride, or a monoacylglycerophosphocholine, from the parent phosphatidylcholine and to characterize them. The following sections will center on these topics.

The choice of the reactions listed above is really dependent on what the investigator wishes to achieve other than formation of sn-glycero-3-phosphocholine. With lithium aluminum hydride, the product would be fatty alcohols, with hydroxylamine, the fatty acyl hydroxamates; with tetrabutylammonium hydroxide, free fatty acids; and finally with sodium hydroxide in methanol, the methyl esters. It is apparent that the latter reaction would give not only the desired glycerophosphocholine but also methyl esters of the fatty acids originally bound to the glycerol backbone. As shall be described later, these esters can be assayed immediately by gas-liquid chromatography coupled with mass spectrometry. The production of the methyl esters and the glycerophosphocholine can be achieved in very facile manner as follows.

Direct Chemical Attack

First, anhydrous sodium hydroxide is dissolved in methanol to make a 0.5 M solution. Then the phosphatidylcholine sample, in chloroform–methanol (1:1, v/v), is placed in a Pyrex glass conical centrifuge tube and evaporated to dryness under a stream of nitrogen. To this residue is added a small volume of chloroform (usually 5–10% of the final volume of methanol). Subsequent to completion perhaps the most effective route to structure proof of phosphatidylcholine is one in which the glycerophosphocholine backbone is obtained in an unaltered form. Essentially this involves the cleavage of the ester bonds and can be achieved by any of the four reaction pathways illustrated in Figure 4-3.

An aliquot of the phosphatidylcholine sample (size will depend on amount of product desired) is evaporated to dryness under nitrogen and dissolved in 0.1 volume of choroform. Sufficient equivalents of 0.5 M NaOH in methanol (prepared by dissolving anhydrous NaOH in 99.5% methanol) are added to allow a two- to threefold excess of that needed for complete cleavage of the ester bonds. The mixture is allowed to stand at room temperature for 15 min, and 6 N HCl is added carefullly with stirring to make the solution slightly acid (pH 6–7). Then equal volumes of choroform and water are added, and the mixture is stirred vigorously using a vortex unit. Two phases will appear, and complete separation can be achieved by centrifugation for 5 min at 500–1000g. The lower phase of which is the chloroform-rich one containing the methyl esters, and the upper phase is the water-rich one containing the glycerophosphocholine. The two phases are carefully separated and the chloroform-rich one is washed with methanol–water (10:9, v/v) until the washes are neutral. These washes can be added to the water-rich component, and the

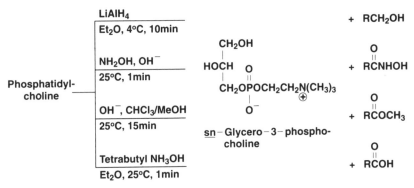

FIGURE 4-3. Reaction pathways for cleavage of the fatty acid esters of phosphatidylcholine.

latter is washed twice with chloroform. There is usually no need to save these chloroform washes.

This base-catalyzed reaction is very smooth and gentle, proceeding to 100% completion in 15 min at room temperature. No isomerization or migration of the polar head group (or formation of a cyclic ester) has been documented.

For our purposes, the characterization of the glycerophosphocholine present in the water-rich phase will have center stage, with a more modest discussion of the fatty acid ester fraction. As a footnote to the above experiment, one should measure carefully the distribution of phosphorus in the two fractions; the phosphorus should be exclusively in the water-soluble fraction.

Glycerophosphocholine: Characterization

On the basis of complete recovery of the phosphorus in the water-rich fraction, there are several analytical techniques that can be used to establish the chemical features of this material.

CADMIUM CHLORIDE COMPLEX. Glycerophosphocholine is obtained as a viscous oil at room temperature, but can be isolated in a crystalline form as the cadmium chloride derivative. In a typical preparation, an aliquot of the water-rich fraction containing 100–500 mg (assumed) of glycerophosphocholine is mixed with an equal volume of 99% ethanol and is reduced to dryness by evaporation *in vacuo* at 35°C. The thick syrup is dissolved in 1 ml of 99% ethanol; and to this clear solution, 500 mg of $CdCl_2(2.5 \ H_2O)$ in 6 ml of water diluted with 99% ethanol to 60 ml, is added slowly with stirring until a slight turbidity is noted. This mixture is allowed to stand at room temperature for several hours during which colorless needles are formed. After storing at 4°C overnight, the crystals are recovered by centrifugation or filtration through Whatman #1 filter paper and are washed twice with cold 99% ethanol and then with diethyl ether. Subsequent to drying *in vacuo* at 65°C, the crystals can be subjected to certain analytical procedures. The yield of the anhydrous derivative is usually in the 80–90% range.

This product has the basic formula of $(C_8H_{20}O_6NP)(CdCl_2)(440)$ (Calculated: P, 7.04; N, 3.18. Found: P, 6.95; N, 3.21). It shows an optical activity value, $[\alpha]_D^{26}$, of $-1.55°$ (in water). Synthetic glycerophosphocholine exhibited an optical activity, $[\alpha]_D^{26}$ of $-1.50°$ (in water). These results would suggest an *sn*-3 configuration for the naturally occurring phosphatidylcholine. However, as discussed before, these optical activity values are very low and one could not prove the presence of a 5–10% contaminant of a different configuration. Nonetheless, the cadmium chloride complex does have merit since it can be easily analyzed for C, H, N, and P, which would support a basic structure; but even more important is that it is very useful in allowing chemical synthesis to the diacyl derivative as will be described.

Further insight into the chemical structure of the glycerophosphocholine can be achieved through use of a few analytical procedures outlined as follows.

VIC-GLYCOL ASSAY.   On the basis of the phosphorus content of the cadmium chloride complex, it would appear that the compound was indeed a glycerophosphocholine. If it possesses an *sn*-3 configuration, then it should have one *vic*-glycol group, $CH_2OH-CHOH-$, and a *vic*-glycol-to-phosphorus molar ratio of 1.0. If it had the *sn*-2 configuration, then there would be no *vic*-glycol group present. The question can be settled through use of the oxidant, periodic acid H_5IO_6, which cleaves *vic*-glycols in a quantitative manner. In this reaction, the periodic acid cleaves the glycol to yield formaldehyde, a phosphorus-containing aldehyde, and the iodate, IO_3^-, ion. The latter is determined by addition of potassium iodide, which is converted to iodine. The latter is titrated with a standard sodium thiosulfate solution in the presence of a starch indicator. A control without a *vic*-glycol present is run since I^- also reacts with periodic acid and the difference in titrimetric values represents periodate consumed. A sorbitol solution of known concentration is used for calibration purposes. A number of other approaches to estimation of *vic*-glycerol content have been published, and particular reference is made to articles by Dixon and Lipkin (1954), Karnovsky and Brumm (1955), and Thompson and Kapoulas (1968). Amounts of a *vic*-glycol-containing compound as small as 10–12 µg can be detected.

Given that this assay shows one mole of *vic*-glycol present per mole of phosphorus, it does not prove whether the stereochemical form is *sn*-3 or *sn*-1. The *sn*-2 form is excluded if the above results are obtained, since 1,3-glycols are not attacked by periodic acid. Perhaps the best route to such a proof is through a biochemical approach, namely, the use of stereospecific phospholipases on the native phosphatidylcholine. However, prior to discussion of this methodology, it is important to complete the basic characterization of the glycerophosphocholine. This will entail the proof of the presence of a glycerophosphoric acid backbone and also the choline moiety.

GLYCEROPHOSPHORIC ACID.   Using glycerophosphocholine as the starting material, it is possible to cleave this compound to yield glycerophosphoric acid and free choline. This can be accomplished in an acid or alkaline media (e.g., 1 N NaOH at 60°C for 4 hr, or 1 N HCl at 60°C for 6 hr) but there is a problem, and this involves the intramolecular migration of the phosphocholine group with resulting isomerization. This can be illustrated by the general equations given in Figure 4-4.

This reaction scheme is highly simplified since an additional isomerization of the *sn*-2 configuration can occur via a cyclic orthoester with *sn*-glycero-1-

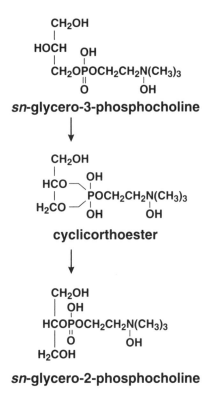

FIGURE 4-4. Intramolecular migration encountered during chemical hydrolysis of glycerophosphocholine.

phosphocholine as the product. Thus, in one reaction system the additional potential for formation of a racemic mixture (sn-1, sn-3) is very high. Subsequent to these isomerizations, release of the choline base can follow. Thus it is patently evident that acid or alkaline hydrolysis is not the method of choice for determination of the stereochemical configuration of the glycero-phosphocholine backbone of naturally occurring phosphatidylcholine.

However, all is not lost since the hydrolytic cleavage of glycerophospho-choline can yield glycerophosphoric acid. Whether it is isomerized or not, it can be converted to barium salt and analyzed. Yields of 90% can be obtained. The resulting barium salt has a molecular formula of $(C_3H_7O_6P)(Ba)(307)$ (calculated phosphorus value, 10.1; found, 10.3).

The comforting facet of the above experimental findings is that now one can show that a glycerophosphoric acid is indeed present in phosphatidyl-choline. Though certain experimental approaches—namely, hydrolysis of phosphatidic acid (i.e., choline-free phosphatidylcholine) in alkaline medium—can produce mainly the unaltered sn-3 form of glycerophosphoric acid, there is still some 5–10% of the sn-2 and sn-1 form present, and this is sufficient to warrant concern in interpretation of the configuration of the material under study.

CHOLINE.   As described earlier, choline is a strongly alkaline compound (pK 13.9); in early studies, it was isolated from an aqueous fraction by treatment with Reinecke's salt (ammonium diamminetetrakis(thiocyanato-N)chromate), phosphotungstic acid, or phosphomolybdic acid. Water-insoluble products are formed; these can be isolated and solubilized (for example, the Reineckate can be dissolved in acetone), and a spectroscopic examination can be conducted. Amounts of choline from 0.3 to 3 mg can be estimated by this approach.

Barak and Tuma (1981) described the assay of choline by use of periodide. A colored complex is formed, and its absorbance is measured in 1,2-dichloroethane. The limit of detection is close to 0.5 μg choline. If phospho-choline is present also, then it must be degraded by a phosphatase to yield free choline.

Another successful approach to determination of choline is through the use of pyrolysis–gas chromatography–mass spectrometry using electron impact (Stavinhoa and Weintraub, 1974). Low nanomolar concentrations of choline can be detected by this sophisticated procedure. A particular advantage of this methodology is twofold: it allows a positive identification of choline by the gas chromatographic pattern of the pyrolysis products, and the mass spec-troscopic peaks provide an absolute proof of structure in this case.

Finally, though only a direct chemical hyrolysis was described above for liberation of choline, it is possible to employ the enzyme phospholipase D to release choline from intact phosphatidylcholine with the concomitant forma-tion of phosphatidic acid.

*Long-Chain Fatty Acyl Groups; Chemical Evaluation
as Methyl Esters*

In the direct chemical attack for proof of structure of phosphatidylcholine (see the section entitled "Glycerophosphochline: Characterization") the decision was made to subject this phospholipid to a base-catalyzed methanolysis. As noted, two products are formed, namely, glycerophosphocholine and the methyl esters of the long-chain fatty acid substituents on the intact phosphatidylcholine. Since the analytical approach to proof of structure of the glycerophosphocholine has been achieved, it is logical now to consider the other product, the methyl esters.

Under the experimental conditions described in the base-catalyzed methanolysis, the yield of methyl esters is essentially quantitative. These esters are easily soluble in chloroform and can be checked for purity by thin-layer chromatography on a silica gel G plate. In a solvent system of petroleum ether–diethyl ether–acetic acid (80:20:1, v/v), the methyl esters will migrate to an R_f of 0.85 and can be detected by using the sulfuric acid/char reaction. Though a very unlikely happening, the presence of any phospholipid P can be excluded by spraying with the phosphorus reagent. If any phospholipid is present, a blue color will develop at the origin (in this solvent system).

It is best to purify the methyl esters by thin-layer chromatography of the sample on another silica gel G (250 μm) plate using this solvent system, but without any spray reagent being used. A comparable plate is run and the methyl ester band is detected by the sulfuric acid/char reaction. Then the unsprayed plate is scraped at the methyl ester area and the silica gel is extracted with petroleum–diethyl ether (80:20, v/v) or with chloroform–methanol–water (1:2:0.8, v/v). It is always prudent to spray the latter plate (after removing the desired area by scraping) with sulfuric acid and then charring. This will validate whether the apparent removal of all of the methyl esters has been accomplished.

The most effective way to analyze quantitatively the composition of the purified methyl esters is by gas-liquid chromatography, using a flame ionization detector. As an alternative, a combination of gas-liquid chromatography and mass spectrometry can provide a wealth of information on composition and structure. The mass of material required for this type of separation depends in part on the size and type of chromatography column and the type of mass spectrometer. Amounts of material as low as 5 μg can be analyzed by this combined technique.

The identification of lipids such as the fatty acids, by gas-liquid chromatography, has been pursued extensively by many different investigators using a variety of sophisticated approaches. The result has been an immense number of publications on this subject. Rather than repeat these very detailed findings, the reader is urged to read an excellent review article by Kuksis (1978). This latter article summarizes the various approaches that can be used to define the composition of a fatty acid ester mixture. An additional publication of value centering on the use of reverse-phase high-pressure liquid chromatography for analysis of fatty acid ester mixtures is one by Arveldano et al. (1983). A particular advantage of the latter methodology is that it can be applied to the determination of specific radioactivities of compounds derived from labeled fatty acids in metabolic studies, but has lesser value for compositional analyses. One drawback is that the method requires microgram quantities of fatty acid esters, whereas in the gas-liquid chromatographic procedure, amounts in the nanogram region can be analyzed using a capillary column.

A Summation and Transition

The basic point of the chemical approach to proof of structure of a naturally occurring phosphatidylcholine was to illustrate the type of information that can be gathered by rather simple, straightforward techniques. The proof of structure is not complete, since the stereochemical purity of the compound and the positioning of the fatty acyl chains—whether saturated and unsaturated are equally distributed between C-1 and C-2 or specifically located on one or the other of the carbons (1 and 2)—remain to be elucidated.

In the following section, emphasis will be placed on establishing the placement of the fatty acyl chains on the glycerol backbone by a combination of enzymological maneuvers and organic synthetic chemistry. At the same time the stereochemical configuration of the phosphatidylcholine will be developed. It is hoped that the reasons for these choices will become self-evident.

A Biochemical Route: Asymmetry and Phospholipases

In the previous section, it was established that the diacylphosphatidylcholine under study contained a glycerophosphocholine backbone, which had characteristics of an *sn*-3 configuration based on the optical activity of its cadmium chloride complex. In addition, the *vic*-glycol analysis showed that there were two hydroxyl functions adjacent to each other on the glycerophosphocholine. Though the fatty acyl substituents on the two hydroxyl functions could be cleaved, recovered as methyl esters, and analyzed by gas-liquid chromatography–mass spectrometry, no decision could be made as to any specific positioning on the parent molecule. At present, there is no chemical reagent (or reaction) that can selectively cleave either the C-1 ester or the C-2 ester linkages on the phosphoglyceride molecule. It is necessary to resort to a biochemical reagent, namely the phospholipase, for resolution of the positioning of the fatty acyl groups and, at the same time, the stereochemical configuration of the native phosphatidylcholine.

Among the phospholipases, phospholipase A_2 has been of major importance in establishing the positional specificity of fatty acids on the glycerophosphocholine backbone. However, two other phospholipases, namely C and D, have been of considerable value in various facets of phospholipid biochemistry. Prior to any further discussion of the positional asymmetry of the fatty acid substituents, it seems appropriate to describe certain features of these phospholipases and how useful they can be in defining the structural features of the phosphoglycerides.

The Phospholipases A_2, C, and D: A Unique Group of Biochemical Reagents

Present knowledge on the phospholipases found in nature can be best summarized in Figure 4-5 as to their action on the phosphoglyceride diacylphosphatidylcholine.

The enzymes noted as phospholipase A_1 and phospholipase A_2 are classified as acyl hydrolases, and those labeled as phospholipase C and phospholipase D are classified as phosphodiesterases. Discussion will be restricted here to these enzymes which are found widely distributed in mammalian cells and which, except for phospholipase A_1, are intimately associated with the signal transduction pathway. There are other enzymes which will attack phospholipids, such as (a) phospholipase B, found mainly in *Penicillium notatum*, which will cleave both the C-1 and C-2 ester bonds, and (b) platelet activating factor acetyl hydrolase, which cleaves the C-2 (acetyl) ester group as well on platelet activating factor (1–*O*-alkyl-2-acetyl-*sn*-glycero-3-phosphocholine). Others are the lysophospholipases, sphingomyelinase, and lecithin-cholesterol acyl transferase (LCAT), which, though very important enzymes, will not be discussed at this point. While the role of the various phospholipases and transferases in cellular signal transduction processes is under intense scrutiny at this point in time, the perspective here will be limited to their application and usefulness in structure proof studies on phosphoglycerides.

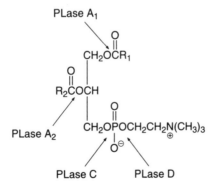

FIGURE 4-5. General mode of attack of phospholipases on a typical substrate, phosphatidylcholine.

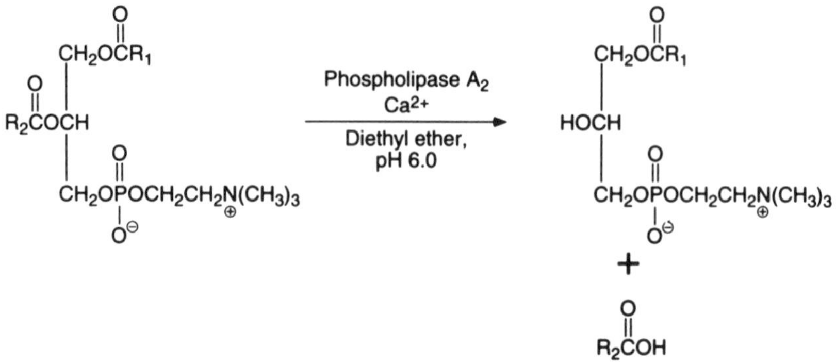

FIGURE 4-6. Phospholipase A_2 attack on diacylphosphatidylcholine.

Phospholipase A_2 (Phosphatide Acyl Hydrolyase,
EC 3.1.1.4)

This enzyme catalyzes the specific hydrolysis of the fatty acid ester located on
the C-2 carbon position of an *sn*-3 phosphoglyceride. The reaction occurs as
shown in Figure 4-6 using 6phosphatidylcholine as the model substrate.

Wells (1971) showed through the addition of oxygen-18-labeled water to
the reaction mixture that the point of attack occurs at the *O*-acyl bond as
shown by the arrows in Figure 4-5. This conclusion was supported by the
finding that oxygen-18 was found only in the liberated fatty acid. Though no
direct evidence for the occurrence of an acyl–enzyme intermediate could be
obtained, formation of such a complex could not be completely excluded.

There has been an enormous amount of information published on the
isolation, purification, and characterization of the phospholipases, particularly phospholipase A_2. Instead of relating a number of these observations,
the reader is strongly encouraged to consult an excellent treatise on this
subject matter by Waite (1987). Consequently only that information pertinent
to the objectives of this section, concerned mainly with the biochemistry of
the venom phospholipase A_2, will be discussed here.

ISOLATION AND PURIFICATION.   While phospholipases A_2 have been isolated from a number of sources (Waite, 1987), only the type obtained from
snake venom will be considered here. The reasons are twofold: First, this
venom contains copious quantities of this phospholipase (and no other type);
and second, this enzyme can be obtained in high purity (and in good yield) by
convenient, easily followed experimental procedures. While it is true that the
crude venom can be used, it is of some concern that the phospholipase A_2
represents only 2–5% of the total protein in the venom, and in the experiments to be described, the other proteins might in some way compromise the
results. Thus, it is preferable to use at least a partially purified enzyme in the

following studies. Certain features of the purification of this enzyme and factors influencing its activity are presented next.

A very satisfactory source of phospholipase A_2 is the venom of the snake, *Crotalus adamanteus* (Eastern diamondback rattlesnake). This venom can be obtained in lyophilized form from commercial suppliers such as Miami Serpentarium (Miami, FL). Of importance, the lyophilization process does not alter the chemical, physical, or enzymatic characteristics of the original venom obtained from this snake.

In an early publication on this topic, Saito and Hanahan (1962) took advantage of the heat stability of the enzyme (resistant to denaturation after heating at 100°C for 10 min) which removed much of the non-phospholipase protein. Then chromatography on DEAE cellulose yielded two proteins with phospholipase A_2 activity in nearly 85% purity. Later, Wells and Hanahan (1969) reported on an improved method for purification of the phospholipase A_2 from this venom. The crude enzyme sample was first suspended in EDTA and passed through a Sephadex G-100 column. All the enzymatic activity emerged in a single peak, and this fraction was further chromatographed on BioRex 70, a weak cation exchange resin, and then on DEAE cellulose. The latter procedure yielded two separate protein peaks with phospholipase A_2 activity. One final step involved passage of these fractions through an SE-Sephadex column which resulted in pure samples. Subsequently, the latter could be obtained in crystalline form. Each exhibited a molecular weight of ~30,000 daltons and had a specific activity of 3200 (as compared to a value of 152 for the starting venom sample). The specific activity is expressed as units per milligram of protein. Units refer to the liberation of 1.0 μequiv of fatty acid in 1 min at room temperature. The two enzymatically active fractions comprised 60% of the original activity and 2–4% of the starting protein.

SUBSTRATE SPECIFICITY. Phospholipase A_2 has been detected in a wide variety of mammalian cells, and its substrate specificity has been studied in considerable detail. Interestingly, in a large number of the phospholipases A_2 found in various cell types, phosphatidylethanolamine is the preferred substrate. However, other phosphoglycerides, such as phosphatidylcholine, phosphatidylinositol (and its phosphorylated derivatives) can be attacked. Recently the isolation and purification of a plasmalogen-specific phospholipase A_2 from platelets and heart tissue has been reported. The occurrence of a selective alkyl ether acylglycerophosphocholine phospholipase A_2 has not been reported to date.

The phospholipases A_2 isolated from snake venom appear able to attack equally well phosphatidylcholine, phosphatidylethanolamine, and other phosphoglycerides, if these compounds are presented in the proper physical configuration. As noted above, the latter can be accomplished through the use of a diethyl ether–methanol–water mixture or by inclusion of a suitable detergent.

CATION REQUIREMENTS.   There is no question that Ca^{2+} is absolutely required for optimal activity of phospholipase A_2. It is intimately involved in the binding of the enzyme to substrate and also in catalytic activity. The molar concentrations needed in these two events vary with the source of the enzyme but in general they are in the 0.1–1.0 mM range (final concentration). A detailed discussion of the influence of Ca^{2+} on the apparent interfacial catalysis by phospholipase A_2 is described in great detail by Scott et al. (1990) and also by Waite (1987).

SOLVENT EFFECTS.   Phospholipase A_2 reactivity is greatly influenced by the physical nature of the lipid aqueous environment—that is, micellar structure of the substrate. The predominant mode of attack is for this enzyme to interact with aggregates of phospholipids. No evidence in support of an interaction of the enzyme with a monomeric form of phospholipid substrate has been presented. Thus reagents such as water, which in effect change dramatically the physical form of the phospholipid substrate, influence the degree of reactivity of phospholipase A_2.

In hindsight, it is now quite clear that the adventitious inclusion of diethyl ether in a phospholipase A_2 reaction mixture (Hanahan, 1952) led to formation of a physical form of the substrate most attractive to the enzyme. Support for this concept was obtained in a very elegant manner by Poon and Wells (1974). Using sedimentation velocity as an indicator, these investigators showed that three hydration states are evident in diethyl ether–water solution of a phospholipid substrate, such as phosphatidylcholine. Data were presented that showed the phospholipases A_2 from *Crotalus adamanteus* venom were active only in a hydration state of 25–30 molecules of water per molecule of phospholipid. Under hydration conditions where there were only seven molecules of water per mole of phospholipid or significantly greater than 30 molecules of water, the enzyme showed no activity. The impact of the addition of water on the phospholipase A_2 activity was shown by Wells and Hanahan (1969; see Figure 1 in this reference). In this study the initial solvent was a mixture of diethyl ether and methanol, 95:5, v/v) which, when the proper amount of water was added, could be an excellent media for phospholipase A_2 activity. The methanol has advantages in that many of the more highly or completely saturated phospholipids are soluble, whereas in diethyl ether alone, their solubility is limited. For more details on the influence of water on the phospholipase A_2-catalyzed reaction in diethyl ether–methanol mixtures, the reader is urged to consult a paper published by Misiorowski and Wells (1974).

As is obvious from the above discussion, the physical state of the phospholipid substrate is very important in any assay involving phospholipases. Besides the ether–methanol–water system, detergents have been used to prepare mixed micelles with the phospholipids. A widely used detergent is the neutral Triton X-100. A more detailed description of this approach can be gained from material presented in a monograph edited by Dennis (1991).

A TYPICAL ASSAY SYSTEM.   A number of different techniques have been used to follow the course of phospholipase A_2 activity toward a phospholipid substrate. A good evaluation of the various methodologies is given by Reynolds et al. (1991). As they emphasize, these assay procedures are more suited for the nanomolar range of substrate and not suitable for evaluating phospholipase A_2 activity in a typical signal transduction experiment since the amounts of free fatty acids that are liberated by an agonist stimulation are not detected by the assays described. Even though a cell may be labeled with radioactive arachidonic acid, and this label incorporated into a phospholipid, its release by introduction of an agonist does not immediately mean that a phospholipase A_2 was involved. There are simply too many other combinations of lipolytic enzymes that could result in liberation of free fatty acid. In studies designed to explore the stereochemical configuration of a phospholipid or to establish the positional specificity of fatty acids on a phospholipid, radiolabeled substrates are of considerable value. These topics will be discussed shortly, but prior to such an event a brief examination of a typical assay system is used to determine the phospholipase A_2 activity of a preparation and is also useful in a study on the positional location of fatty acids of a naturally occurring phospholipid. This methodology can be used also in ascertaining the stereochemical characteristics.

In a typical reaction, 10–15 mg of phosphatidylcholine (egg lecithin is a typical substrate) is dissolved in 2–2.5 ml of diethyl ether–methanol (95:5, v/v) and transferred to a 5-ml glass-stoppered volumetric flask. To this clear solution is added 20 µl of an aqueous solution containing 20 mM Ca^{2+} and approximately 3 µg of purified phospholipase A_2. The flask is stoppered and the contents are mixed well for 4–5 min and then allowed to stand at room temperature for 30–45 min. Usually a precipitate forms during the course of the reaction; but as noted before, this does not compromise the extent of the reaction.

If any analytical evaluation of the progress of the reaction is wanted, then an aliquot can be diluted with 95% ethanol (or methanol) and titrated with 0.02 N methanolic NaOH with cresol red as the endpoint indicator. An alternate approach is to employ a titrimeter (Radiometer, microtitration assembly, Copenhagen), but it is important to realize that most phospholipases A_2 adhere strongly to glass surfaces. Thus a vessel resistant to the solvents used in the reaction is mandatory, as is rigorous attention to cleaning the electrode after titration. The extent of the enzymatic reaction can be calculated using as a control a reaction mixture with no enzyme added.

If an examination of the chemical nature of the products is desired, then an aliquot of the reaction mixture can be subjected to thin-layer chromatography on silica gel G plates. In a solvent system of chloroform–methanol–water (65:35:7, v/v), the liberated fatty acids will migrate to an R_f near 0.90, the unreacted substrate to an R_f near 0.42, and the lysophosphatidylcholine to an R_f near 0.18. These compounds can be visualized by spraying the plate with TNS reagent (see Chapter 3) and then exposing to ultraviolet light. The desired spots are removed by scraping and are then extracted with solvent, usually chloroform–methanol–water (1:2:0.8, v/v); their composition is determined by any number of analytical procedures (P assay, fatty acid composition, FAB and GC/MS spectral analysis, and, if sufficient compound, optical rotatory behavior).

STEREOSPECIFICITY AND POSITIONAL SPECIFICITY OF PHOSPHOLIPASE A$_2$. It is now abundantly clear that the phospholipases A$_2$ obtained from mammalian cells and from snake venoms share similar general characteristics. In particular, they will attack only the *sn*-2 acylester position on a phosphatidylcholine, invoking a positional specificity. Concomitantly, this attack will occur only if the substrate has an *sn*-3 configuration. There is no detectable attack on an *sn*-1 form, and only 50% attack on a phosphoglyceride having an *sn*-2 configuration. Thus, a stereochemical specificity is invoked. Finally, though not widely appreciated, the stereochemical conformation of the phosphorus atom on the polar head group (of phosphatidylcholine) influences the degree or extent of phospholipase A$_2$ action. Thus, a chiral effect is evoked.

As mentioned in an earlier section, there are *no* chemical reagents that can establish with any certainty the specific position of attachment of an acyl ester on a diacylphosphatidylcholine or any other phosphoglyceride. The same is true in any attempt to establish the stereochemical conformation of phosphatidylcholine, whether *sn*-1 or *sn*-3, by strictly chemical means. The solution to this dilemma is to use synthetic phosphoglycerides of defined structure as well as phospholipase A$_2$ to establish the correct stereospecificity of the latter enzyme. Certainly, this is reminiscent of the "chicken or egg" argument, yet it does work as will be explained below.

Synthetic Schemes.   There are basically two main routes to preparation of "mixed acid" (enantiomeric) diacylphosphatidylcholines. The first involves a total *de novo* synthesis and the second centers on a partial or semisynthesis using a highly purified naturally occurring phosphatidylcholine as starting material. Both approaches have merits and drawbacks, which are noted as follows.

Total De Novo Synthesis.   Although the organic chemical techniques used in the *de novo* synthesis are most provocative and can be categorized as works of (chemical) art, they do present a challenge. These are not "trivial pursuits," but they have been most important in establishing the stereochemical conformation of naturally occurring phosphoglycerides, the positional location of specific fatty acyl groups on these molecules, and the specificity of phospholipase A$_2$ action. Despite the difficulty of the undertaking, it is possible to synthesize a phosphatidylcholine with a different fatty acyl residue on the *sn*-1 and *sn*-2 hydroxyl functions. This can be accomplished by the reaction scheme which is shown in Figure 4-7 (short form only).

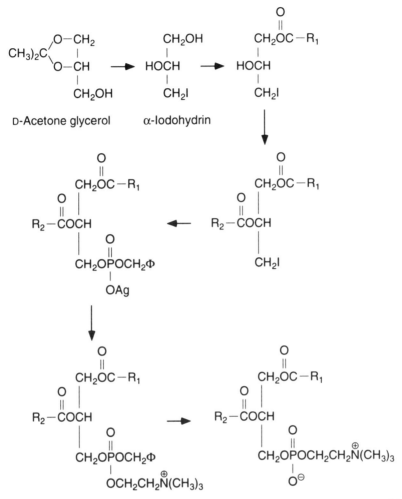

FIGURE 4-7. Total *de novo* chemical synthesis of *sn*-3 diacylphosphatidylcholine with different fatty acyl groups on the *sn*-1 and *sn*-2 positions.

In this reaction sequence, a different fatty acyl group can be substituted at the *sn*-1 position of the optically active iodohydrin, and then subsequently a different fatty acyl residue can be substituted at the *sn*-2 position. This derivative is then converted by a series of steps to the desired phosphatidylcholine form. If one's mind is set on a *de novo* synthesis, then reference should be made to the classic experimental paper by de Haas and van Deenen (1961) (and also earlier publications) and an equally impressive review article on phospholipid synthesis by Eibl (1980).

Over the past 15 years or more, the importance of phospholipids in biological reactions has stimulated commercial suppliers to market several different synthetic forms of phosphatidylcholine and other phosphoglycerides. Hence, both *sn*-1 and *sn*-3 diacyl (single type) phosphatidylcholines are available in addition to a wide variety of mixed acyl (different fatty acyl groups on the *sn*-1 and *sn*-2 positions) phosphatidylcholines. Racemic (*sn*-1 plus *sn*-3) phosphatidylcholine is also available. Each of these compounds will be discussed below in the context of phospholipase A₂ specificity of action and the total structure of naturally occurring phosphoglycerides.

Partial (Semi) Synthesis.   A blend of two methodologies can be used quite effectively to prepare mixed acid phosphatidylcholines, which are of prime importance in elucidating the structure of naturally occurring phosphoglyceride and phospholipase A₂ activity. Perhaps the best scenario would be to illustrate the activity of the phospholipase A₂ (from *Crotalus adamanteus* snake venom) toward a racemic phosphatidylcholine sample and toward individual *sn*-1 and *sn*-3 enantiomers. In each case the same result would be found, and so only the racemic mixture reaction is depicted in Figure 4-8.

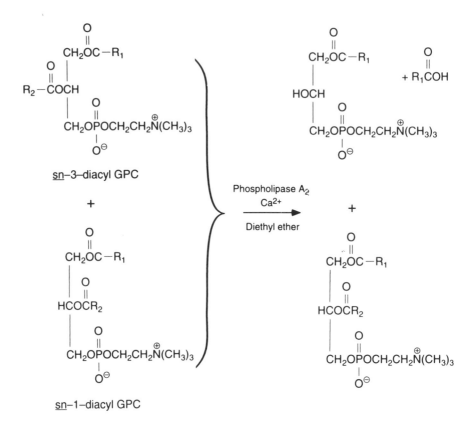

FIGURE 4-8. Behavior of snake venom phospholipase A₂ on racemic phosphatidylcholine.

This reaction is run under the same conditions as described earlier and the products isolated by preparative thin-layer chromatography. The products should be analyzed for P content (which would be primarily monoacylphosphatidylcholine and unreacted *sn*-1 diacylphosphatidylcholine) and for optical activity. Whereas the starting *rac*-diacylphosphatidylcholine would exhibit no optical activity, each of the above products of the reaction should have optical activity. The liberated free fatty acid can be converted to a methyl ester form and examined by gas–liquid chromatography coupled with mass spectrometry (GC-MS).

A Reservation. Inasmuch as the starting material was *rac*-dipalmitoyl phosphatidylcholine, the data from the above experimental protocol tells only that the enzyme has stereochemical preference for the *sn*-3 enantiomer. Even though it is easy to show that there was an equimolar release of free fatty acid and formation of a lysophosphatidylcholine, it is *not* possible to tell whether the enzymatic attack occurred at the *sn*-1 or the *sn*-2 acyl ester bond. In any event, the stereospecificity of phospholipase A₂ has been established by this experimental approach, and with this information a route to proof of specific positioning of fatty acyl substituents on naturally occurring phosphoglycerides is accessible.

Positional Asymmetry of Fatty Acyl Groups. The previous exercise showed that the phospholipase A₂ is a potent chemical reagent in effecting the formation of a specified stereochemical form—namely the *sn*-3 lysophosphatidylcholine—from a *rac*-phosphatidylcholine mixture. A similar result can be achieved by synthesis of well-defined phosphoglycerides by starting with pure *sn*-glycero-3-phosphocholine which can be produced very smoothly and quantitatively from a phosphatidylcholine (isolated from chicken egg) as outlined in Figure 4-9. This procedure was developed by Brockerhoff and Yurkowski (1965). Using their reaction conditions, 3 g of diacylphosphatidylcholine, incubated in 30 ml of diethyl ether plus 3 ml of 1.0 M methanolic tetrabutylammonium hydroxide for 1 hr at room temperature, will yield a

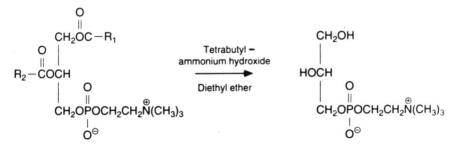

FIGURE 4-9. Chemical preparation of *sn*-glycero-3-phosphocholine (GPC).

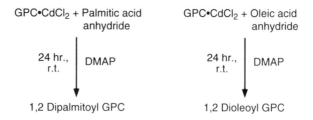

where: GPC is <u>sn</u>-glycero-3-phosphocholine and
DMAP represents N,N-dimethyl-4-aminopyridine

FIGURE 4-10. Chemical synthesis of 1,2-Dipalmitoyl-*sn*-3-GPC and 1,2-Dioleoyl-*sn*-3 GPC.

glassy precipitate of *sn*-glycero-3-phosphocholine. The latter residue is dissolved in methanol and then reprecipitated with diethyl ether. Pure *sn*-glycero-3-phosphocholine is recovered in an 80–85% yield. Its optical rotation value and elemental composition was as expected.

This pure glycerophosphocholine can be used as starting material for the synthesis of mixed acid phosphatidylcholine as shown in Figure 4-10.

1,2-Dipalmitoyl-*sn*-3 GPC and 1,2-dioleoyl*sn*-3 GPC. This reaction is based on a method developed by Gupta et al. (1977) in which fatty acid anhydrides are the acylating agents and *N,N*-dimethyl-4-aminopyridine is the catalyst.

1-*O*-palmitoyl-2-*O*-oleoyl-*sn*-glycero-3-phosphocholine and 1-*O*-oleoyl-2-*O*-palmitoyl-*sn*-glycero-3-phosphocholine. The availability of the dipalmitoyl and dioleoyl derivatives of *sn*-3-GPC, as prepared in Figure 4-10, above makes it possible simply to subject each derivative to the action of phospholipase A_2. This results in the formation of the corresponding lyso derivatives, 1-*O*-palmitoyl-2-hydroxy(lyso)-*sn*-glycero-3-phosphocholine and 1-*O*-oleoyl-2-hydroxy-*sn*-glycero-3-phosphocholine. The conditions for this enzymatic cleavage have been described previously. Then these derivatives can be reacylated, using the conditions described earlier and a different fatty acyl group introduced in each case. This is illustrated by the equations in Figure 4-11.

Usefulness of the Above Synthetic Products. The mixed acid phosphatidylcholines obtained by the above synthetic procedures can be used to support (in part) the specificity of positioning of the fatty acyl residues on these phosphoglycerides and the specificity of attack by phospholipase A_2. As a test of these two propositions, incubation of these two species of "mixed acid" phosphoglycerides with phospholipase A_2 will yield the products given in Figure 4-12.

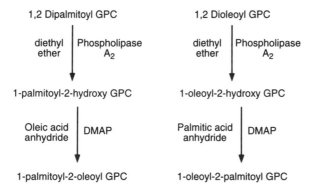

FIGURE 4-11. Preparation of phosphatidylcholine with dissimilar fatty acyl groups on the *sn*-1 and *sn*-2 positions.

Conclusions: Both Right and Wrong. Basically these results support a specific attack of the enzyme on the *sn*-2 ester bond, and certainly the release of a specific fatty acid would support such a conclusion. Earlier proof that the phospholipase A_2 prefers the *sn*-3 configuration and will not attack the *sn*-1 form allows one to be quite comfortable in a decision on its stereospecificity. Yet, no where here has it been proven unequivocally that the partial synthesis outlined in Figure 4-11 above does give the fatty acids substituted in the described positions. The only route to such proof is through the use of *de novo* synthesis as discussed above. This approach will allow one to make claims as to the position of fatty acids on phosphoglycerides.

An Addenda: Stereospecificity of Phospholipase A_2. Attack on Phosphoglycerides Chiral at Phosphorus. Only within the past 15 years has any attention been paid to the influence of the phosphorus atom in the catalytic capability of phospholipase A_2 on a typical phosphoglyceride such as phosphatidylcholine. However, our current understanding of the influence of the phosphorus configuration on phospholipase A_2 activity has derived largely from observations made in the laboratory of Tsai and colleagues, who used chirally modified substrates. In this brief description of these exciting advances, the phosphothioate derivatives will be considered.

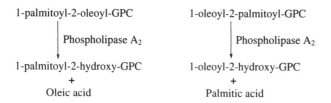

FIGURE 4-12. Phospholipase A_2 action on a "mixed acid" phosphatidylcholine.

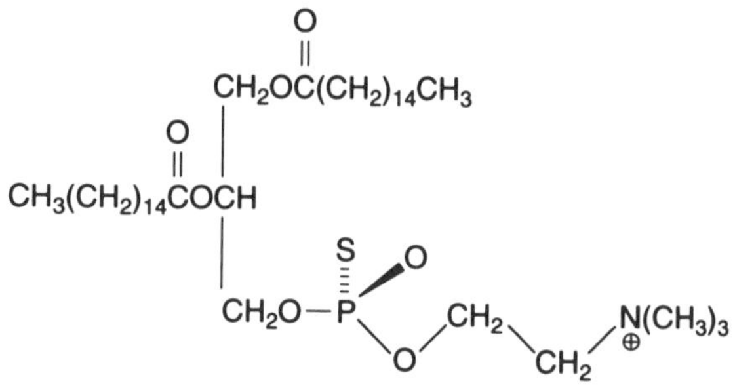

(R$_p$)–DPP$_S$C

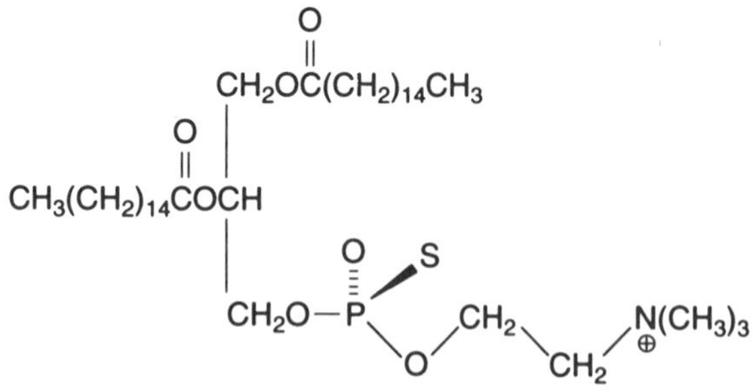

(S$_p$)–DPP$_S$C

FIGURE 4-13. Examples of chirally substituted phosphatidylcholines.

Perhaps the best approach to this subject is to give two examples of chirally substituted phosphatidylcholines (Figure 4-13).

The term DPP$_S$C represents 1,2-dipalmitoyl-sn-glycero-3-thiophosphocholine. (R)$_P$ and (S)$_P$ (using the nomenclature system of Cahn, Ingold, and Prelog; see Chapter 1) refer to the diastereoisomeric forms which are distinguished from each other by ^{32}P nuclear magnetic resonance. These derivatives can be synthesized (as a mixture) by the method outlined by Bruzik et al. (1983). The absolute configuration of the chiral thiophosphoglycerides was established by Jiang et al. (1984).

A most provocative observation by Bruzik et al. (1983) using a mixture of the diastereoisomers (R$_P$)-DPP$_S$C and (S$_P$)-DPP$_S$C was the ability of the enzyme, phospholipase A$_2$, to distinguish between the isomers. This was quite unexpected since the P atom, which is five bonds removed from the susceptible carboxyl function, should have little influence on the course of the reaction. In any event, this reaction sequence can be described in the following

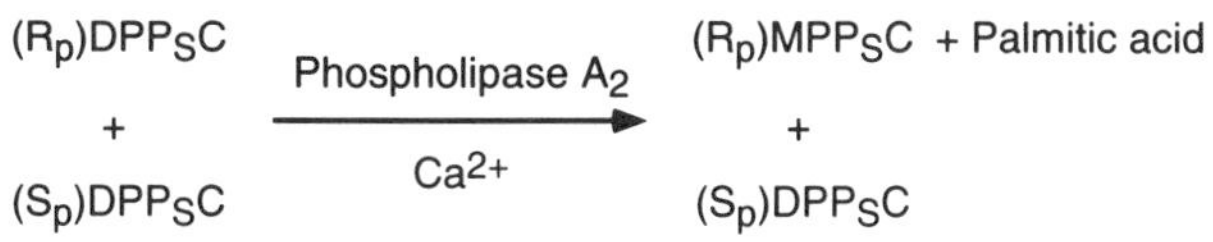

$(R_p)DPP_SC$ Phospholipase A_2 $(R_p)MPP_SC$ + Palmitic acid

+ $\xrightarrow{\qquad Ca^{2+} \qquad}$ +

$(S_p)DPP_SC$ $(S_p)DPP_SC$

FIGURE 4-14. Phospholipase A_2 action toward two chirally substituted phosphatidylcholines.

abbreviated form illustrated in Figure 4-14. The reaction products can be isolated quite easily by thin-layer chromatography and further characterized. Essentially all sources of phospholipase A_2 examined to date have shown a preference for the R_p isomer of the thio-substituted phosphatidylcholine. The same preference was exhibited toward the R_p isomer of thio substituted phosphatidylethanolamine.

Two questions can be raised regarding the use of these phosphothioates in studies with phospholipase A_2. The first is the low total reactivity of phospholipase A_2 toward DPP_SC as compared to DPPC (without the thio substituent), which is only 4% relative to the natural isomer. The second question relates to the possible chirality of the phosphorus in a membrane phosphoglyceride. To date, the first question appears not to be of immediate importance since the course of the reaction is similar in the thio and nonthio derivatives. As regards the second question, it has not been possible to establish with confidence the configuration of the phosphorus in membrane-bound phosphoglycerides.

The main point in this section is to illustrate that the configuration of the polar head group can infuence the reactivity of phospholipase A_2. Conceivably this same effect could translate to membrane biochemical pathways involving phosphoglycerides. If further details on the phosphothioates are desired, the review article by Bruzik and Tsai (1991) is recommended.

If additional information on the chemistry of phosphorus is desired, two books, the first of which is authored by Goldwhite (1981) and the second of which is edited by Quin and Verkade (1981), are recommended reading.

Phospholipase C (E.C. 3.1.4.3)

This enzyme, classified as a phosphodiesterase, catalyzes the hydrolysis of the ester bond between the diglyceride and the polar head group of a phosphoglyceride, such as phosphatidylcholine. This reaction is outlined in Figure 4-15.

Essentially the existence of this enzymatic activity was first established with certainty in the early 1940s in the filtrates of the bacterial organism *Cl. perfringens* (*Cl. welchii*) and has been detected subsequently in the filtrates of many other bacteria. Sloane-Stanley (1953) first presented evidence supportive of the presence of a phospholipase C in a mammalian cell, the guinea pig brain. Since that time, momentum in this field of study has increased exponentially with the exciting developments in the signal transduction field. In the latter system, stimulation of a cell leads to activation of a phospholipase C,

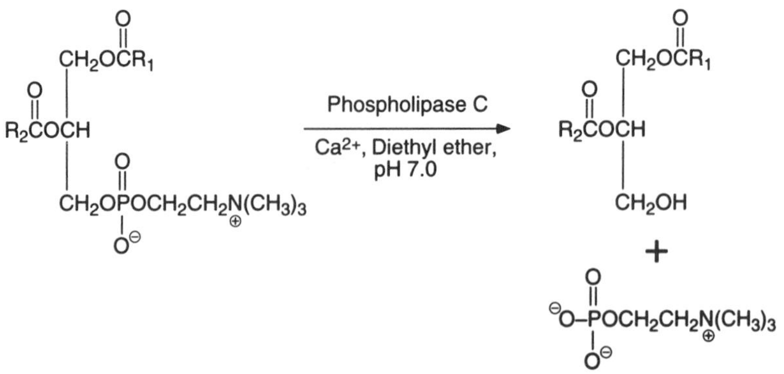

FIGURE 4-15. Mode of attack of phospholipase C on diacylphosphatidylcholine.

which cleaves a substrate such as phosphatidylinositol-4,5-bisphosphate to yield a diglyceride and inositoltriphosphate. These two derivatives have significant effects on protein kinase C and on Ca^{2+} movement from intracellular stores, respectively. Phospholipase C has been found in many different mammalian cells.

ISOLATION AND PURIFICATION.   A considerable amount of literature has accumulated on this subject in the past several years, and an excellent summary on this topic can be obtained from an article by Rhee et al. (1991). Basically, these investigators found that there were four isoforms of phospholipase C present in bovine brain. These can be isolated by a combination of ion-exchange, heparin–agarose, and reversed-phase chromatography. Interestingly, though their molecular weights and other physical characteristics were different, all of these isoforms had similar catalytic capabilities.

A thoughtful and detailed outline of the isolation, purification and characterization of the bacterial phospholipases C is given by Waite (1987). Procedures similar to those mentioned above for the mammalian enzymes were also useful in defining the chemistry of the bacterial phospholipases C.

SUBSTRATE SPECIFICITY.   Some very interesting observations have been made on the substrate preference exhibited by the bacterial and mammalian derived phospholipase C. There are some unique specificities encountered, which are described as follows.

A considerable amount of data has been reported on the substrate preference of the phospholipase C present in the organism *Bacillus cereus*. Interestingly, three phospholipases C have been isolated and purified, the first of which has high specificity for phosphatidylcholine, the second for phosphatidylinositol, and the third for sphingomyelin (often termed sphingomyelinase). Similar substrate requirements have been noted in the phospholipase C isolated from other bacteria.

In a study designed to investigate the structural features of a phospho-glyceride interaction with a bacterial phospholipase C, El-Sayed et al. (1985), reported that the carbonyl group and its closely related environment are most important. A more detailed treatment of the substrate specificity of this en-zyme can be found in an excellent review by Massing and Eibl (1994).

When attention is directed toward the phospholipase C found in mammalian tissue, a rather unique and different substrate profile is evident. It appears that the "most favored substrate status" must be assigned to the inositol- containing phosphoglycerides, namely, phosphatidylinositol (PI), phosphatidylinositol phosphate (PIP), and phosphatidylinositol-4,5-bisphosphate (PIP_2). There is some evidence that the plasma membrane of certain mammalian cells contains a phospholipase C with high specificity for the bisphosphate, PIP_2. The latter enzymatic interaction would be closely associated with the signal transduction pathway in mammalian cells.

Current dogma states that the preferred substrate for mammalian phospho-lipase C is a phosphoinositide on which arachidonic acid is esterified at the sn-2 position—for example, PIP_2. However, Holub and Celi (1984), pre-sented evidence showing that mammalian phospholipase C is highly specific toward inositol-containing phosphoglycerides, but not with respect to the fatty acid composition of the phosphoglyceride. Rather, in the intact cell, specific release of an arachidonate-rich diglyceride by phospholipase C action may relate more to compartmentation of the substrate than to enzyme preference.

Cation Requirements.   While some of the phospholipases C found in bac-teria appear to prefer Ca^{2+}, there are many many reports supporting Zn^{2+} as the divalent cation of choice. There is some support for the fact that this enzyme is probably a metallo (Zn^{2+}) protein which also requires Ca^{2+} for catalytic activity, but there is more evidence for the enzyme's ability to influ-ence the surface charge on the micellar substrate system.

In contrast to the phospholipase C of bacterial origin, there appears to be a consensus that the enzyme from mammalian sources prefers Ca^{2+} at millimo-lar concentrations. Interestingly, heavy metal ions such as Hg^{2+} or Zn^{2+} are strong inhibitors of this source of enzyme and EGTA must be included in the reaction mixture to chelate these cations.

OTHER ACTIVATORS.   Only limited attention has been paid to the influence of solvents on phospholipase C activity, However, they do exert a positive effect on the reaction—in particular, diethyl ether (as diethyl ether–ethanol mix-ture). This was especially true for the catalysis of phosphatidylcholine cleav-age by the phospholipase C from *Cl perfringens*.

In many of the systems described in the literature for phospholipase C assay, there is a decided preference for inclusion of a detergent—for example, deoxycholate or Triton X-100. In the detergents or in the diethyl ether–ethanol, it is very likely that these reagents cause a change in the size (and or configuration) of the substrate to one suitable for preferred interaction with the enzyme.

STEREOSPECIFICITY OF ATTACK.   In a particularly definitive review on the substrate specificity of phospholipase C isolated from *Bacillus cereus,* Massing and Eibl (1994) showed that this enzyme attacked preferentially a phosphatidylcholine containing an *sn*-3 configuration and was inactive toward an *sn*-1 and *sn*-2 form as well as toward lysophosphatidylcholine (1-*O*-acyl-2-hydroxy-*sn*-glycero-3-phosphocholine). Also it exhibited no activity toward sphingomyelin. The behavior of a sphingomyelin-specific phospholipase (a sphingomyelinase) will be discussed later in this chapter. It is presumed that comparable phospholipases in mammalian cells will act in the same way.

TYPICAL ASSAY SYSTEM.   Structural proof on a phosphoglyceride, such as diacylphosphatidylcholine, can be accomplished quite smoothly through the use of phospholipase C. Inasmuch as this enzyme liberates specifically a chloroform-soluble diacylglycerol (diglyceride) and a water soluble *O*-phosphocholine in excellent yields, each of these fragments can be isolated and subjected to certain analytical techniques. This enzymatic assay can be conducted as desribed below and in essence follows the procedures outlined by Satouchi and Saito (1979) and by Satouchi et al. (1981).

A sample of diacylphosphatidylcholine containing 50–150 μg phospholipid P is dissolved in 5 ml peroxide-free diethyl ether. If the primary intent is to obtain the diglyceride only, then the reaction is run in 0.1 M borate buffer, pH 8.0. On the other hand, if the phosphorylated bases are the main objective, then 0.01 M Tris maleate buffer, pH 7.95, is preferred. In each protocol, however, 2 μmol of NaCl, 22 μmol of $CaCl_2$ and approximately 2.5 mg of enzyme protein (phospholipase C from *Bacillus cereus*, Calbiochem-Behring Corp., La Jolla, CA) are added. The two-phased system is stirred vigorously (in a hood) for 2–3 hr with a Teflon stir bar. Progress of the reaction is monitored by removal of an aliquot of the ether-soluble phase and subjecting it to thin-layer chromatography on a silica gel G 250-μm plate in a solvent system of chloroform–methanol–water (65:35:7, v/v). Use of a phosphorus spray and a subsequent char reaction will show the extent of the enzymatic reaction. When the starting substrate has been exhausted, a stream of nitrogen is directed on the reaction mixture (placed in the hood) and the diethyl ether is removed. The aqueous suspension is then mixed with sufficient chloroform and methanol to make a 1:2:0.8, (v/v) mixture. To this clear solution is added 0.5 volume each of chloroform and water and the combination then mixed by vigorous shaking.

The water-rich upper phase is removed and assayed for organic phosphorus. This will provide a quantitative evaluation of the efficiency of the enzymatic cleavage. The chloroform-rich lower layer is washed well with methanol–water (10:9, v/v). An aliquot of this chloroform-soluble fraction can be tested qualitatively on a small silica gel G plate for the presence of phospholipid P. If reaction conditions were chosen correctly, then this test should be negative.

The chloroform-soluble fraction should contain only sn-1,2-diacylglycerol (though a small amount of sn-1,3-diacylglycerol might be present (due to intramolecular migration of the sn-2 acyl group to the sn-3 position), and it can be converted to the t-butyldimethylchlorosilyl derivative as outlined by Satouchi and Saito (1979). Using a combination of gas-liquid chromatography and mass spectrometry, the composition and structure of this diacylglycerol can be accomplished. If the presence of 1,3-diacylglycerol is suspected, it can be detected by the thin-layer chromatographic system described by Matsuzawa and Hostetler (1980). In this procedure, a "double" undimensional thin-layer chromatography is conducted on silica gel H plates impregnated with 0.01 M magnesium acetate. Using 10-cm $\times$ 20-cm plates, for example, the samples are applied to the plates in the usual way and the first solvent, chloroform–methanol–water (65:35:5, v/v), is allowed to migrate to approximately 7 cm from the origin. The plate is removed from the tank and dried under nitrogen. The plate is then placed in a solvent of heptane–diethyl ether–formic acid (90:60:4, v/v) which is allowed to run to the top of the plate. After drying, a spray—for example, TNS—can be applied to locate the lipids. The following (approximate) R_f values were reported for the indicated lipids: free fatty acids, 0.73; 1,2-diacylglycerol, 0.64; 1,3-diacylglycerol, 0.60; monoacylglycerol, 0.48; phosphatidylcholine, 0.32; lysophosphatidylcholine, 0.18. Standards can be run to verify the structure assignment on a sample under study.

The water-soluble fraction (from the original extraction of the reaction mixture described above) is evaporated to dryness, dissolved in 2 ml water, and applied to a Dowex AG 50 W $\times$ 8 column (H form, 200–400 mesh-column size, 0.6 $\times$ 5 cm). Water is used as the eluant, and the P-containing components are collected. The latter could be evaporated to dryness and silylated using the reagent pyridine/Tri-Sil "TBT" and BSTFA (1:2:2, v/v) (Pierce Chemical Co., Rockford, IL). Subsequently an aliquot of this TMS derivative can be analyzed directly by the gas-liquid chromatography technique of Karlsson (1970). For further information on the detection and assay of choline and phosphocholine, suggested reading is an article by Kennerly (1991).

As mentioned earlier, many investigators have used detergents instead of diethyl ether in the phospholipase C assay; however, these compounds are to be avoided if the intention is to analyze the phosphorylated bases liberated in this reaction. Interference in the separation and assay procedures is a problem.

Phospholipase D (EC 3.1.4.4)

This enzyme is categorized as a phosphodiesterase, but it also has trans-phosphatidylation (transfer) activity as well. As a phosphodiesterase it can be a useful reagent for use in structure proof studies on diacylphosphatidylcho-

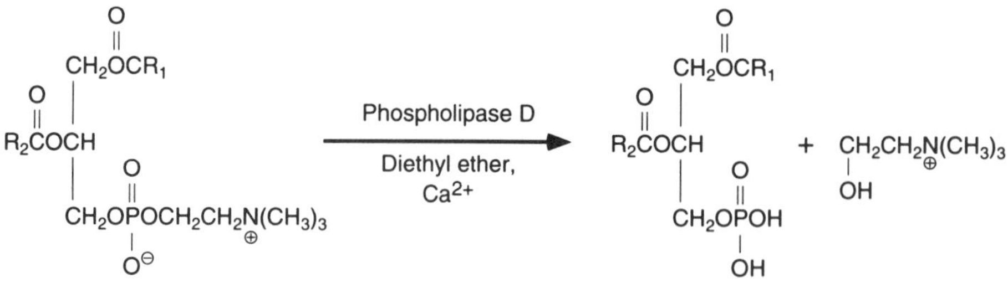

FIGURE 4-16. The phosphodiesterase activity of phospholipase D on a phosphatidyl-choline.

line as well as on other phosphoglycerides. This enzyme catalyzes release of the base, choline, from phosphatidylcholine as shown in Figure 4-16. Diethyl ether can exert a powerful effect on the catalytic rate.

Proof for the existence of phospholipase D in nature (i.e., in plant tissues) was provided by Hanahan and Chaikoff (1947). Subsequently, this enzyme has been detected in microorganism and mammalian cells. An overall, in-depth treatment on phospholipase D is given by Waite (1987).

In addition to its capability as a phosphodiesterase, this enzyme also was found by Yang et al. (1967) to possess transphosphatidylase activity (essentially a base exchange reaction) as illustrated in Figure 4-17.

In this example, including as much as 10% ethanol in the reaction mixture, phospholipase D will catalyze the incorporation of ethanol in place of choline, yielding a phosphatidylethanol derivative.

ISOLATION AND PURIFICATION.   A number of approaches have been published on the purification of phospholipase D activity, ranging from use of rat

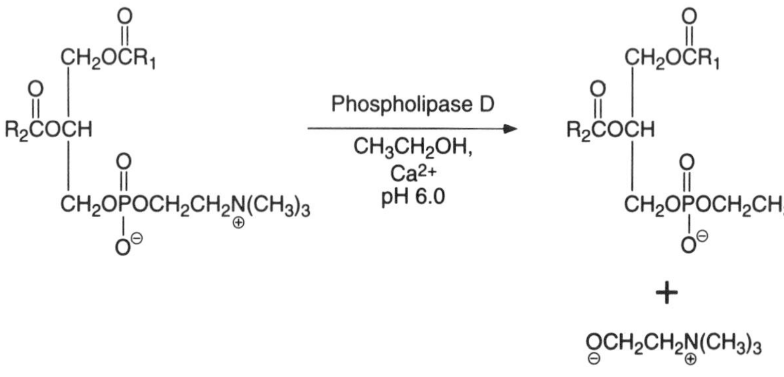

FIGURE 4-17. The unique transphosphatidylase activity of phospholipase D on a phosphatidylcholine.

brain tissue (Kobayashi and Kanfer, 1991) to use of plant tissues [i.e., cabbage leaves (Allyger and Wells, 1979)]. All of these sources have both phosphodiesterase as well as transphosphatidylase activity. An unusual phospholipase D, found in mammalian blood serum, possesses an activity specific only for glycosylphosphatidylinositol, an important component of the (GPI) anchor of many membrane proteins (Huang et al., 1991). For the purpose of general convenience only, an abbreviated form of the procedure for isolation of phospholipase D from cabbage leaves, as described by Allyger and Wells (1979) will be summarized here.

An acetone powder of Savoy cabbage leaves (prepared by heat coagulation and acetone precipitation and available from commercial sources) is suspended in diethyl ether and then filtered. The residue is dispersed in sodium acetate and calcium chloride at pH 5.0, and the insoluble matter is separated by centrifugation at 13,000 g for 10 min at 4°C. The supernatant contains all the phospholipase D activity and in our laboratory's experience can be used successfully in any experiment requiring this enzymatic activity. It can be maintained at -20°C for 1–2 weeks.

If a further purification is desired, then the supernatant (above) can be subjected to ammonium sulfate fractionation, gel filtration on Sephadex G-200, and finally on α-aminopropane-agarose affinity column. On polyacrylamide gel electrophoresis in the presence of sodium dodecylsulfate, the final product migrated as a single band with an estimated molecular weight of 113,000. Upon sedimentation equilibrium velocity ultracentrifugation, an estimated molecular weight value of near 117,000 was obtained.

SUBSTRATE SPECIFICITY.   Inasmuch as there are two enzymatic activities present in the phospholipase D preparation, it is necessary to address each one separately:

Phosphodiesterase (Hydrolysis) Activity.   A rather broad substrate specificity is exhibited by the purified phospholipase D (phosphodiesterase activity). It can attack phosphatidylcholine, phosphatidylethanolamine, phosphatidylserine, and phosphatidylglycerol. In most cases, Ca^{2+} was an activator, but variable results were obtained on the positive influence of diethyl ether on the catalytic activity of different sources of this enzyme. Usually the optimum pH was in the range from 5.0 to 7.0. Mammalian phospholipase D, containing both the phosphodiesterase and transphosphatidylase activities, exhibited a broad-range substrate specificity similar to that of the plant enzyme. However, the mammalian enzyme showed a dependency for the presence of oleic acid in the reaction system (Kobayashi and Kanfer, 1991).

Transphosphatidylase (Transfer) Activity.   This catalytic activity shows a particularly broad substrate specificity with as many as 20–30 acceptors (all with a free hydroxyl function) being effective in the reaction. Ca^{2+} and diethyl ether both have positive effects on the catalytic rate of the enzyme activity. The pH range of the reactions tends to be from 6.0 to 8.0. In the transphosphatidylase reaction system, increasing amounts of the acceptor molecules will shift the reaction away, but not completely from the hydrolytic function. A particularly effective discussion of the use of the phospholipase D transphosphatidylase activity in the synthesis of a variety of phospholipids from a single starting compound such as phosphatidylcholine is well presented in an article by Eibl and Kovatchev (1981). To those uninitiated in the vagaries and difficulties in a strictly organic chemical sythesis of enantiomeric phosphoglycerides, the usefulness of phospholipase D to accomplish certain of these syntheses by a smooth, uncomplicated, and high-yield reaction cannot be overemphasized.

STEREOSPECIFICITY.   This area of research has not been pursued to date with the same intensity as previously described for phospholipase A_2. In large part this can be attributed to the lack of highly purified (or even crystalline) phospholipase D. Nonetheless, this problem has been addressed to a limited degree, and the results are encouraging. Bugaut et al. (1985), who also investigated the stereospecificity of phospholipase C in the same paper, have used a commercial preparation of phospholipase D (from cabbage) to study its behavior toward (*rac*)-1-lauroyl-2-oleoyl-glycero-3-phosphocholine and 3-lauroyl-2-oleoyl-glycero-1-phosphocholine. They found that the initial rate of hydrolysis by this enzyme toward the *sn*-1 and *sn*-3 isomers was in a ratio of 40–50:1. These results showed that phospholipase D exhibited a stereopreference rather than a stereospecificity for this substrate. Interestingly, substitution of the 2-acyl group with an alkyl residue led to complete loss of stereopreference of phospholipase D (but not of phospholipase C). Presumably the carbonyl oxygen on the 2-acyl substituent is important to the recognition of enantiomers by phospholipase D. When an alkyl group is present instead (of the acyl group), then spatial reference is lost.

In another detailed study on the influence of the substituent at the C-1 position in *sn*-3 choline phosphoglycerides, Waku and Nakazawa (1972) investigated the catalytic behavior of the phospholipase D of cabbage and carrot. Their findings showed that either source exhibited high activity toward a diacylphosphoglyceride, with a much lower rate of hydrolysis toward an alkylacyl- and alkenylacylphosphoglyceride.

It appears on the basis of the above studies that phospholipase D does exhibit a stereopreference and also has preference for a phosphoglyceride with a particular substituent on the C-1 position. However, further definition of the behavior of phospholipase D requires the use of highly purified enzyme.

ACTIVATORS.   Both Ca^{2+} and diethyl ether are recognized as important in stimulating reaction catalyzed by phospholipase D. Most probably the calcium ions interact with the enzyme protein, whereas the diethyl ether undoubtedly influences the physical structure of the substrate.

ASSAY SYSTEM.   Two approaches will be described, with the first being the use of this enzyme for structure proof purposes and the second being the involvement of this enzyme in signal transduction processes in cells.

Structure Proof.   The high specificity of phospholipase D toward the P–O–N base bond in a molecule such as diacylphosphatidylcholine allows proof of the location of the nitrogen base in this and comparable molecules. In a typical reaction, 100 mg (pure) phosphatidylcholine is dissolved in 5–7 ml of peroxide-free diethyl ether, and to this solution is added 10 mg of phospholipase D (cabbage, grade B, Cal Biochemicals, San Diego, CA) dispersed in a buffer containing 1.0 M sodium acetate and 0.2 M $CaCl_2$. This mixture is then stirred vigorously (using a Teflon bar) at room temperature (in a chemical hood) for 8–10 hr. Usually the reaction is complete in this time period. It is possible to use a detergent such as deoxycholate to make a homogeneous mixture, and the reaction will work well but no better than the diethyl ether system. In addition, there are problems with removal of the detergent and recovery of the product, phosphatidic acid.

In the work-up of the diethyl ether reaction, the solvent is removed (in a hood) under a stream of nitrogen. Then the ether-free aqueous dispersion is mixed with 10 ml of a 5 mM EDTA solution to bind the calcium ions. Without the latter treatment, the calcium salt of the phosphatidic acid cannot be recovered (easily) from the reaction mixture. Thus, subsequent to the EDTA treatment, the sample is mixed with sufficient chloroform and methanol to make a mixture of chloroform–methanol–water (1:2:0.8, v/v). Then 0.5 volume chloroform and 0.5 volume water are added, and the mixture is vigorously shaken for 1 min and then allowed to separate into two phases. The upper, water-rich phase contains free nitrogen base—in this case, choline. It can be analyzed by the techniques described earlier.

The lower, chloroform-rich phase is separated carefully from the protein-containing interface, and then it is washed twice with methanol–water (10:9, v/v) and the washes are discarded. The chloroform layer contains the phosphatidic acid (as a sodium salt) and can be isolated by acetone precipitation. The yields can be of the order of 90–95%. One alternative route to identification of the chloroform-soluble material is to analyze it for total phosphorus and total fatty acid ester (see procedures described earlier). In the case of diacylphosphatidylcholine as the substrate, the fatty acid ester/P molar ratio should be 2.0. Another approach is to subject the chloroform-soluble fraction to preparative thin-layer chromatography on silica gel H (calcium ion free) in a two-dimensional system with a solvent system of chloroform–methanol–28% ammonium hydroxide (65:35:6, v/v) in the first direction and a solvent system of chloroform–acetone–methanol–glacial acetic acid–water (4.5:2:1:1.3:0.5, v/v) in the second direction. The phosphatidic acid will not migrate far in the basic solvent (R_f 0.10) and will show an R_f value one-half of that of any remaining starting substrate (R_f 0.40) in the second solvent. Of course with a simple substrate system, one can use the basic solvent in one dimension only and isolate the product from the plate as described earlier. The product, phosphatidic acid, can be analyzed by FAB–mass spectrometry. A mass ion $[MH]^+$ and a $[MH-H_3PO_4]^+$ ion should be easily identifiable.

Phospholipases in Signal Transduction

There is overwhelming evidence in support of the involvement of the above phospholipases—A2, C, and D—in the signal transduction pathway in mammalian cells. However, this topic is beyond the intended scope of this book, and the reader is urged to consult a detailed review on this subject by Nishizuka (1992). Nonetheless, it is appropriate to discuss in brief how the presence of these enzymes in an intact cell subjected to an agonist stimulation is determined. Again the human platelet will be the target cell and the agonist will be platelet activating factor (1-O-alkyl-2-O acetyl-sn-glycero-3-phosphocholine), PAF. The following protocols are typical of those in common use to assess the activity of these phospholipases in stimulated cells.

PHOSPHOLIPASE A2.   In the usual experimental situation, the platelets are prelabeled with tritiated arachidonic acid. Then subsequent to stimulation with PAF, the release of free (unesterified) arachidonic acid is measured.

Labeling of Cells.   Human platelets (1.25×10^9 cells per milliliter) are suspended in a Tyrode-gelatin buffer (pH 6.5) and to this mixture is added 1 μCi of 5,6,8,9,11,12,14,15-[³H]arachidonic acid (90–100Ci per mmol) dissolved in ethanolic bovine serum albumin (BSA) solution (0.1% ethanol and 0.0005% BSA). This mixture is then incubated for 30 min at 37°C. Then the platelets are mixed with 0.5% BSA and 0.1 mM EGTA and centrifuged at 800 g for 15 min. The platelet pellet is washed with a Tyrode-gelatin buffer (pH 6.5; containing 0.1 mM EGTA) twice and then suspended in Tyrode-gelatin buffer (pH 7.2) containing 0.1 mM EGTA to a final cell count of 6×10^8 cells per milliliter. Under these experimental conditions, the tritiated arachidonic acid is incorporated primarily into phosphoglycerides such as phosphatidylcholine, phosphatidylethanolamine, and phosphatidyl inositol-4,5-bisphosphate (PIP_2).

Measurement of Arachidonic Acid Release.   The prelabeled platelet preparation above is preincubated with 10 μM indomethacin (added to prevent conversion of arachidonic acid to prostanoids) for 1–2 min at 37°C and with 1 mM $CaCl_2$ for an additional 1 min. Then platelet activating factor, usually at 1 $\times$ 10^{-8} to 1 $\times$ 10^{-9} M (final concentration), dispersed in 0.005% BSA, is added and the incubation is allowed to continue for 2 min. At this point, sufficient chloroform and methanol are added to make a final concentration of chloroform–methanol–water of 1:2:0.8, v/v. Then 0.5 ml of chloroform and 0.5 ml of 0.5 mM EGTA in 2M KCl are added. The entire mixture is subjected to vigorous shaking, and the two phases are allowed to separate, and the lower phase is recovered and evaporated to dryness under a stream of nitrogen. The residue is dissolved chloroform, previously saturated with water and applied to a precoated silica gel G thin-layer plate. The plate is developed in a solvent system of petroleum ether–diethyl ether–acetic acid (50:50:1, v/v) for neutral lipids; another comparable plate is developed in chloroform:methanol:acetic acid:water (75:45:12:3, v/v) for phospholipids. Individual components are identified by their mobility compared with that of authentic standards (run in a separate lane). The radioactive species can be located by examination of the entire plate with a BioScan unit (BioScan System 200 Imaging Scanner, BioScan, Washington, D.C.). Subsequently, specific radioactive areas can be removed by scraping into a vial, and a quantitative assessment of the labeled components can be made by liquid scintillation counting.

With regard to the phospholipase A_2 protocol, stimulation of human platelets (prelabeled with tritiated arachidonic acid) with the agonist PAF leads to the release of only 5–6% of the arachidonic acid in the free (unesterified) form.

The remainder of the radioactivity is associated with three main phospholipid classes, namely, phosphatidylcholine, phosphatidylethanolamine, and the phosphoinositides, mainly PIP_2. Nonetheless, this type of result is commonly attributed to the action of a phospholipase A_2, which is activated upon agonist interaction with the platelet. The other presumed product, a lysolecithin, would not be labeled in the above experimental protocol and thus, due to the very small amount formed in the reaction, could not be detected. Though one could potentially label the polar head group of the parent phosphoglycerides, there is little need to do so since the arachidonic acid is associated almost exclusively with the *sn*-2 ester position on phosphoglycerides. Consequently, the release of free arachidonic acid can be safely attributed to phospholipase A_2 activity. While the yield of arachidonic acid is very low, the activation of the cell occurs only over a short time span, anywhere from 5 sec to 1 min. Thus self-control of cell activation is evident.

PHOSPHOLIPASE C.   The primary target of phospholipase C action in the platelet centers on the phosphoinositides, primarily PIP_2. The products are diacylglycerol and inositol-1,4,5-tris phosphate. The overall experimental protocol is similar to that described for phospholipase A_2 above, but with a different radiolabel.

Labeling of Platelets.   Two possible routes to labeling of these cells present themselves one of which is to incubate the platelets with myo-[^{3}H]inositol (50 μCi/ml) for 3 hr at 37°C and the other of which is to incubate platelets with [^{32}P]phosphate (150–200 μCi/ml) for 1 hr at 37°C. In each case, approximately 3×10^8 platelets per milliliter are used.

Measurement of Phosphoinositide Turnover.   As a result of these labeling experiments, two analytical procedures can be used to assess the involvement of the phosphoinositides in stimulus-induced activation of platelets by the agonist, PAF. One uses the measurement of the change in phosphoinositide levels per se as determined by [^{32}P]phosphate labeling, and the other uses an assay to determine the accumulation of [^{3}H]inositol phosphates through myo-[^{3}H]inositol labeling.

[^{32}P]Phosphoinositide Level.   Essentially this will show the effect of the agonist, PAF, on the phosphoinositide levels in human platelets. In a typical experiment, PAF is added at 1×10^{-9} M (final concentration) to the labeled platelets together with 1 mM Ca^{2+} (final concentration). After incubation of this reaction system at 37°C for periods of time from 0 to 15 min, the reaction is stopped by the addition of sufficient chloroform, methanol, and 2.4 M HCl to make a 1:2:0.8 (v/v) mixture, followed by vigorous mixing. Then sufficient chloroform is added to make a two-phase system, and again vigorous mixing is initiated. The lower, chloroform-rich layer is recovered and evaporated to dryness under a stream of nitrogen. The residue is dissolved in 50–100 μl of chloroform–methanol (2:1, v/v) and applied to silica gel HL plates (Analtech, Newark, DE). Subsequently the plate is placed in a solvent system of chloroform–methanol–20% aqueous methylamine (30:18:5, v/v). After development in this solvent, the plate can be autoradiographed using Kodak XAR film and the radioactive areas can be identified. In addition, the plates can be sprayed with 2-*p*-toluidinylnaphthalene-β-sulfonic acid (TNS) reagent and individual components can be visualized under ultraviolet irradiation. Individual components are located by reference to the migration of authentic standards—for example, PIP$_2$, PIP, PI, and PA. Consequently on the basis of the radioactive areas detected as described above and using the R_f values of the standards, the comparable areas in the reaction system lanes can be scraped into vials and subjected to liquid scintillation counting for a more accurate evaluation of the levels of products.

In experiments with PAF as the agonist, this reaction would show the level of PIP$_2$ to increase nearly 135% compared to a control within 1 min and would also show the PIP level to increase to approximately 225% of the control in 3 min. Changes in the PI level were not significant. Of importance, phosphatidic acid levels rose very rapidly within 30 sec to a value approximately eightfold greater than that of the control.

[^{3}H]Inositolphosphate Labeling.   Platelets labeled with myo-[^{3}H]inositol phosphates are incubated in Tyrode-gelatin buffer (pH 7.2) containing 1 mM Ca^{2+} and 10 mM Li^+ (all final concentrations) for 10 min. This concentration of Li^+ has been shown to inhibit the conversion of inositolphosphates to inositol. As described earlier, the cell preparation is stimulated with 1×10^{-9} M PAF and at the desired time intervals the reaction is stopped by the addition of 5% trichloracetic acid (w/w) and the mixture is centrifuged at 900 *g* for 5

min. The supernatant is removed, washed with water-saturated diethyl ether (to remove any remaining lipid), lyophilized, and stored in a freezer until ready for assay. At this point, the lyophilized sample is dissolved in 250 μl of distilled water and filtered (0.45-μ Millipore-type HV filter), and the filtrate is then applied to a 25 × 0.4 cm Partisil 10 SAX column in an HPLC unit. The A_{254} value of the HPLC effluent are measured and fractions are collected every 30 sec. The radioactivity in each fraction is monitored by liquid scintillation counting. The various inositol phosphates are identified by spiking the cell extracts prior to the HPLC step.

In this experimental protocol, the results for the formation of tritiated phosphate inositol are as follows (based on percent of control): InP, 450; InP$_2$, 360; InP$_3$, 270. These results support the conclusion that PAF stimulates the breakdown of phosphoinositides via a phospholipase-C-mediated pathway.

PHOSPHOLIPASE D. The presumed endpoint of phospholipase D activity in signal transduction is the formation of a phosphatidic acid by phosphodiesterase cleavage of the polar head group (base) from phosphoglycerides. However, the finding of phosphatidic acid in a stimulus induced activation of a cell may be misleading indicator for phospholipase D activity. The reason is simply that phosphatidic acid could derive from phosphorylation of diacylglycerol formed by phospholipase C action. If one were to try a different approach and use the formation of free choline as an indicator of phospholipase D activity, then this assumption could again prove to be suspect. In this case, phospholipase C can liberate O-phosphocholine from phosphatidylcholine in certain cells, which then could be cleaved by a phosphocholine phosphatase to yield free choline. Thus, this poses a dilemma.

Perhaps the most effective method for assaying for phospholipase D activity in stimulated cells is to take advantage of its transphosphatidylase activity. It has been shown that (mammalian) cell phospholipase D possesses both the hydrolase activity and the ability to transfer a phosphatidyl group to an acceptor molecule. In general, the time course and the agonist requirements are similar for both activities. As a result, a currently popular assay technique has evolved in which the transphosphatidylase activity is employed.

Experimentally, the cells are incubated in the presence of 0.5–1.0% ethanol together with agonist, with the endpoint being the isolation of phosphatidylethanol. The formation of the latter compound is consistent with the presence of a phospholipase D activity.

This experiment is usually run using [^{3}H]ethanol and locating the radioactive product by thin-layer chromatography. An alternative method would be to label the cells with [^{3}H]oleic acid and [^{32}P]phosphate and then monitor the ratio of the tritium label to phosphorus-32 label in any detectable phosphatidic acid and compare to the parent phosphoglycerides. The ratio would not change if the phosphatidic acid were derived by action of phospholipase D in the stimulated cells. However, the assay of choice at the present time is that associated with the formation of phosphatidylethanol.

Specifically, human platelets (1.25 to 2.5 × 10⁸ cells/ml), prepared as described earlier are incubated with 100 μCi [³H]ethanol (s.a. 15–30 Ci/mmol) and PAF (1 × 10⁻⁸ M) for up to 2 hr and at 37°C. The cells are recovered by centrifugation, washed, and then subjected to chloroform–methanol extraction. The chloroform-soluble fraction is then subjected to thin-layer chromatography on silica gel 60 in a solvent system of chloroform–pyridine–formic acid (50:30:7, v/v). In this solvent system, phosphatidylethanol migrates to an R_f of 0.60, slightly but distinctly above that of phosphatidic acid at R_f 0.55. These compounds can be detected by examination of the developed plate with a BioScan 200 Imaging System (Bioscan, Washington, D.C.). The radioactive areas are compared to authentic standards, run in a separate lane on the same plate, and then scraped into a vial for liquid scintillation counting. Finally, if sufficient material is produced, it can be assayed by FAB-MS, in which an *m/e* 125 ion is diagnostic for *O*-phosphoethanol.

SUMMARY STATEMENT ON PHOSPHOLIPASES.   Attention has been centered here on three phospholipases, A_2, C, and D. The weighting of the topic was intentional since these three phospholipases are useful in structural proof studies on phospholipids, and in addition, have a primary role in signal transduction in cells. Notwithstanding this skewing phenomena, there are other phospholipases of importance found in many tissues and cells. Examples are (a) phospholipase A_1, an enzyme which cleaves the acyl ester bond at the C-1 position of diacylphosphoglycerides, thereby liberating a free fatty acid and the corresponding glycerophosphobase; and (b) phospholipase B, an enzyme which attacks, with different velocities, both the C-1 and C-2 ester bonds of a diacylphosphoglyceride. The enzyme initially found in *Penicillium notatum* has been studied *in extenso* by Saito et al. (1991). In addition, an enzyme with similar properties has been found in guinea pig intestine by Gassama-Diagne et al. (1989). The latter authors prefer to call the intestinal enzyme a glycerol ester lipase with broad specificity. The enzymes from both of these sources have been purified to a high degree, and partial characterization has been reported. Interestingly, phospholipase A_1 appears to lack a stereopreference for a phosphoglyceride substrate and appears not to be implicated to any degree in the signal transduction process. Another interesting enzyme is sphingomyelinase, which exhibits a high specificity toward sphingomyelin with liberation of *N*-acylsphingosine (ceramide) and *O*-phosphocholine. Finally, the lysophospholipases are an interesting group of enzymes capable of attacking lysophosphoglycerides with liberation of a free fatty acid and the corresponding glycerophosphobase.

At the present rate of experimental activity in this field, there is no doubt that the importance of these latter enzymes in metabolic processes will be established within the next few years. At present, though, these enzymes are of limited value to structure proof studies on phospholipids. If a more rigorous description of the characteristics of these enzymes is required, reference should be made to the book by Waite (1987) and to individual articles in the review edited by Dennis (1991).

Ether Linked Choline Phosphoglycerides: A Complicating Factor

In the material presented up to this point in this chapter, the impression was given that phosphatidylcholine was the compound of ultimate choice in biological membranes. However, life is never that simple; and as reflected by the heterogeneity of the cells' membrane lipids, this is a true statement. In most mammalian cells, the so-called phosphatidylcholine fraction is really a mixture of the diacyl–monoalkylmononacyl–monoalkenymonoacyl derivatives; although the diacyl component may be the predominant form, the ether-linked phosphoglycerides are also important components. In the main, there are two structural forms of the ether-containing phospholipids as represented in Figure 4-18.

The chemical structure on the top in Figure 4-18 represents 1-O-alkyl-2-O-acyl-sn-glycero-3-phosphocholine (saturated ether type). In this formulation, n represents 15 or 17 while x is usually 14 or 16 with one or two double bonds present. The compound on the bottom represents 1-O-alkenyl-2-O-acyl-sn-glycero-3-phosphocholine (vinyl ether form). In this case, n is equal to 13–15 while x usually is 14 or 16 with one or two double bonds present. Similar-type structures are present in mammalian cells as the eth-

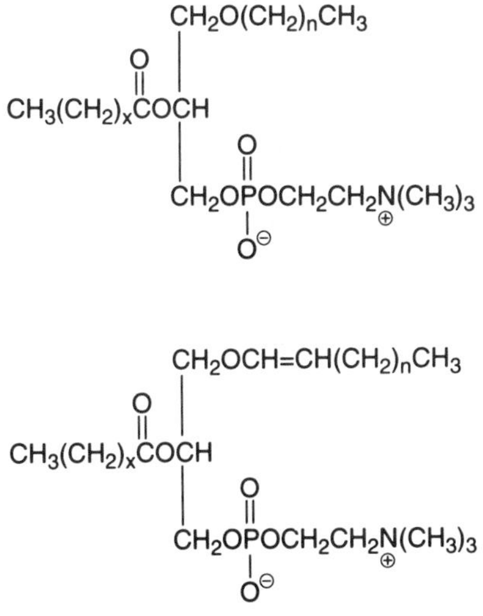

$$CH_2O(CH_2)_nCH_3$$
$$CH_3(CH_2)_xCOCH$$
$$CH_2OPOCH_2CH_2N(CH_3)_3$$

$$CH_2OCH=CH(CH_2)_nCH_3$$
$$CH_3(CH_2)_xCOCH$$
$$CH_2OPOCH_2CH_2N(CH_3)_3$$

FIGURE 4-18. Chemical structures of naturally occurring ether-linked phosphoglycerides. See text for further description.

anolamine derivative (in place of the choline group). These compounds will be discussed in the following chapter.

As mentioned in Chapter 2, the alkylacylcholine phosphoglycerides predominate in the choline-containing fraction in the majority of mammalian cells. On the other hand, the alkenylacyl form predominates in the ethanolamine-containing fraction. It is important to stress that these ether-linked lipids are found essentially only in the choline- and ethanolamine-containing phosphoglyceride. The reason for this high specificity of distribution is not known at present.

Some Brief Comments on Ethers in Biological Systems

The ether bond —C—O—C— is widespread in nature. Among the nonlipid types, two general examples will suffice, namely, thyroxine and guiacol (1-methoxyphenol) There are many others that could be mentioned, but the main emphasis here is on the ether-linked lipids, which are represented by the alkylacyl- and alkenylacyl types. If one wishes a more detailed discussion of the basic chemistry of the ether bond, the excellent book edited by Patel (1967), is recommended reading. If one's interest is more directed toward the chemistry and biochemistry of the ether-linked lipids, then reference should be made to books edited by Snyder (1972) and by Mangold and Paltauf (1983). Even though these references may sound like ancient literature, the information contained in these books would be of paramount importance to anyone launching an investigation in this area.

An Approach to Proof of Structure of Ether-linked Phosphoglycerides

At the present time there is no satisfactory route to isolation of diacyl-, alkylacyl-, or alkenylacylcholine phosphoglycerides as individual components—for example, only the diacyl type or the alkylacyl form, and so on. In general, a differential analytical approach must be invoked and this will be discussed later. However, since diacylphosphatidylcholine has been discussed, proof of structure of ether-linked choline phosphoglycerides as individual species will be described as well as some of their structural features. Later, as the big picture emerges, the route to analysis of a naturally occurring mixture will be examined.

Alkylacylphosphatidylcholine

The basic assumption at this point is that a highly purified preparation of this compound has been obtained from a mammalian cell. Essentially, then, what is needed to prove the structure of this compound? It is hoped that the following sections will show clearly some of the avenues available for pursuit.

General Analytical Procedures

Exactly the same methodology as applied to analysis of diacylphosphatidyl-choline can be used here. Thin-layer chromatography using several different types of adsorbents will show that usually the alkylacyl derivative will co-migrate with the diacyl—as well as the alkenylacyl—phosphatidylcholine counterparts.

Analysis for phosphorus and nitrogen should give a molar ratio of 1:1 and a phosphorus value near 4% (on a dry weight basis). At the same time, an assay for choline should reveal a choline-to-phosphorus molar ratio of 1.0. In each case, exactly the same methodology as described earlier in this chapter for the diacylphosphatidylcholine is applicable to the alkylacyl form.

Specific Structure Proof Procedures Using Chemical and Biochemical Reagents

As noted above, an assumption is made that a purified alkylacylcholine phos-phoglyceride has been isolated free of any diacyl or alkenylacyl form. The order of the experiments described below is not critical. Most of these proce-dures can be adapted for use at the macro- or microlevel.

BASE-CATALYZED METHANOLYSIS. This reaction has been discussed ear-lier; but in the case of the alkylacyl derivative, one of the products is quite different, as shown in Figure 4-19.

In the usual reaction, a 10 mg ($\sim$400 μg of phosphorus) sample is dis-solved in 0.1 ml of chloroform, 2 ml of methanolic KOH (0.5 M) is added, and the mixture is allowed to incubate at room temperature for 10–15 min. Sufficient 6 N HCl is added to neutralize the base, and chloroform plus water are added in sufficient quantity to make a chloroform–methanol–water mix-

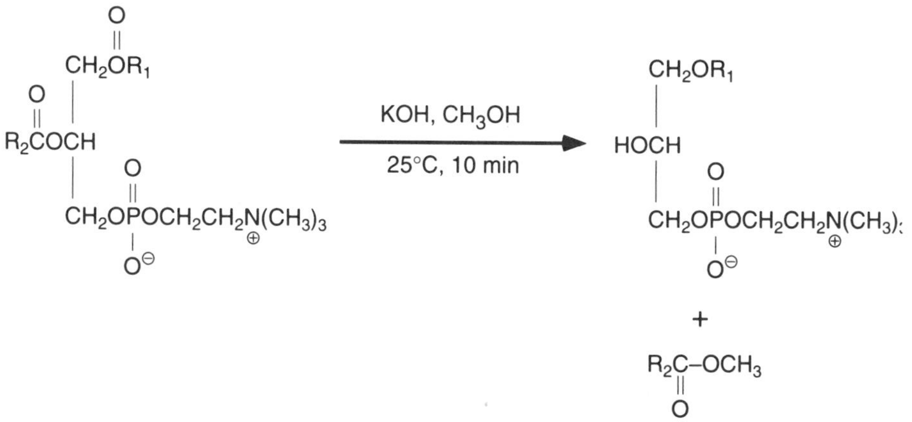

FIGURE 4-19. Base-catalyzed methanolysis of alkylacyl GPC.

ture of 1:2:0.8, v/v). To this clear solution are added 0.5 ml of chloroform and 0.5 ml of water, and the mixture is vigorously stirred. The lower, chloroform-rich layer is recovered and washed twice with methanol–water (10:9, v/v). This extract should contain all the originally added phosphorus since both the monoalkylglycerophosphocholine and the fatty acid methyl ester are chloroform-soluble. A simple phosphorus assay will tell how successful the extraction has been. Recoveries are usually in the 95–99% range.

The question remains as to identification of the products and of their yields. This can be answered very simply by transferring a sufficiently large aliquot of the chloroform-soluble fraction to a silica gel G-coated thin-layer chromatography plate. The latter is developed in a solvent system of chloroform–methanol–water (65:35:7, v/v). The products can be located by spraying with TNS reagent alone if subsequent isolation is desired and/or with phosphorus spray followed by charring with sulfuric acid for qualitative identification. The TNS spray should show two spots, if the reactions went to completion, at R_f 0.85 for methyl ester of a long-chain fatty acid and R_f 0.18 for the alkyl-glycerophosphocholine. If any residual starting material is present, it will be detected at R_f 0.40.

The thin-layer chromatography profile alone will provide a quite accurate evaluation of the extent of the cleavage reaction. If a more quantitative assay is desired, then another plate—usually a preparative type—can be run and individual components can be identified by TNS spray, removed by scraping, and extracted with chloroform–methanol–water (1:2:0.8, v/v) and analyzed after phasing the desired compound into chloroform. Then, the yield of the phosphorus-containing compound can be determined. The methyl esters can be assayed by GC-MS. The presumed alkylglycerophosphocholine then can be studied further.

The results of this reaction show that there must be a fatty acid ester bond in the original compound and that in addition there is another "lipidlike" component, one containing phosphorus and choline (i.e., a monoalkylglycero phosphocholine).

The following procedures can be conducted on the monoalkylglycerophos-phocholine obtained in the section entitled "Base-Catalyzed Methyanolysis" above or on the original alkylacylphosphatidylcholine.

ACETOLYSIS.   The reaction involves subjecting the sample to treatment with acetic acid anhydride and glacial acetic acid at 150°C as outlined in Figure 4-20. The experimental protocol is very simple. A sample containing 20–25 μg phospholipid P is mixed with 0.5 to 1.0 ml of glacial acetic acid–acetic anyhdride mixture (3:2, v/v) and heated at 150°C in a sealed tube for 5 hr. The reaction tube is cooled to room temperature, the seal broken, and the contents evaporated to dryness under a stream of nitrogen. The residue is dissolved in 0.5 ml of chloroform–methanol (1:1, v/v) and examined by thin-layer chro-matography on a 250-μm silica gel G plate in a solvent system of petroleum ether (b.p. 30–60°C)–diethyl ether (v/v). The diacetate should migrate to an R_f value of 0.85.

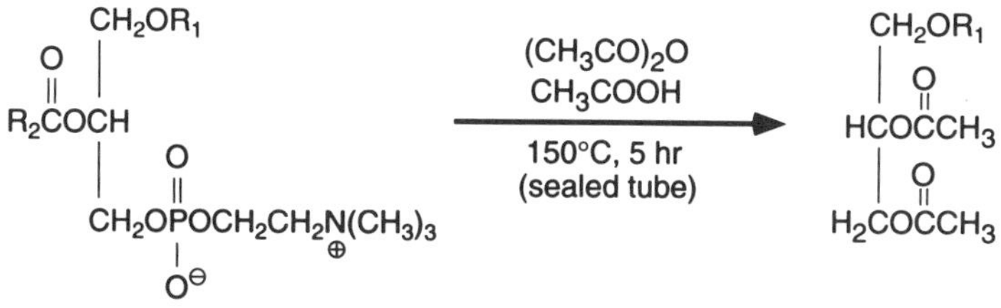

FIGURE 4-20. Acetolysis of alkylacyl GPC.

A further preparative chromatogram can be run, and the desired area can be identified by TNS spray and removed by scraping. A mixture of chloroform/methanol/water (1:2:8, v/v) can be used to extract the diacetate from these scrapings. Subsequent to recovering the derivative into a chloroform-rich phase, it can be analyzed by GC alone or GC-MS by comparing the profiles to those of synthetic, well-defined standards. This acetolysis technique is quite satisfactory on samples as small as 400 μg (total weight), with excellent results.

HYDROGENOLYSIS.   Vitride, which is chemically identified as sodium bis(2-methoxyethoxy)aluminum hydride, is a very effective reducing agent. It is easy to handle and can attack the alkylacyl GPC or its lysoderivative to yield the free alkylglycerol (glyceryl ether) as illustrated in Figure 4-21. The glyceryl ethers can be isolated by thin-layer chromatography and subjected to the following identification procedures:

Diacetate Derivative, Isopropylidene Formation, and Mass Spectrometry. A particularly facile route to proof of structure of the glyceryl ether components is to use a combination of (a) purification by thin-layer chromatography and (b) separation into individual species of the isopropylidene or acetate derivatives by gas-liquid chromatography. As an example, the experimental protocol for examination of the diacetates will be described at this point and later that of the isopropylidene derivative. Inherent in such an experimental approach is to have synthetic standards of high purity available. It is possible to accomplish this task, but it is not necessarily quick and neat. The synthesis

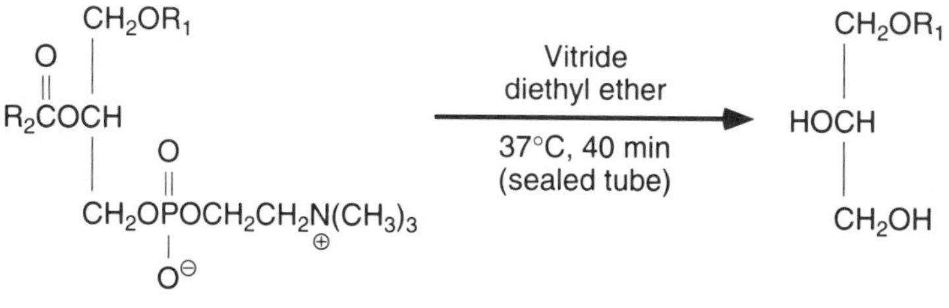

FIGURE 4-21. Vitride cleavage of alkylacyl GPC.

of enantiomeric glyceryl ethers and their derivatives are outlined by Chacko and Hanahan (1968) and by Paltauf (1983). These are not simple procedures, but given care and diligence, yields in the range of 35–40% can be expected. The experimental description that follows is based on the isolation of the glyceryl ethers by the methods cited earlier. In this instance, the analysis will be done with the diacetate derivatives. A gas chromatograph equipped with a flame ionization detector (a Varian or a Hewlett-Packard are very suitable), together with a data collection system, is used. Gas chromatographic separation of the acetylated compounds can be accomplished with a 6-ft × 4-mm-i.d. glass column packed with 10% SP-2399 on 100/200 Chromosorb WAW. The column is operated isothermally at 210°C with detector and injector temperature at 230°C. A helium flow of 30 ml/min is maintained throughout the run. Standard synthetic glyceryl ether diacetates are run, and their retention times are compared with the unknown glyceryl ether sample for tentative identification. If further identification is needed, the effluent from another sample run on the column described earlier is fed into a Finnegan MAT Model III mass spectrometer attached to the gas–liquid chromatography unit. In this latter case, it is best to undertake the separation on a 20-meter SE54 glass capillary column using helium carrier gas. The ion source is maintained at 250°C, the accelerating voltage is 3 kV, the electron energy is 70 eV, and the cathode current is 1 mA. Further details of these experimental procedures are described by Kumar et al. (1984).

A significant amount of information can be obtained in such mass spectrometric examinations, but only a limited excursion into this area will be made here. The electron impact mass spectrum of 1-O-hexadecyl-2,3-diacetyl glycerol would give the following peaks of importance.

Fragment	m/z	Interpretation
M-(60 + 43)	297	Results from loss of an acetyl group and acetic acid from the 16:0 compound (m/z, 325 for 18:0 derivative)
$CH_2O(CH_2)_{15}CH_3$	255	Cleavage of the bond between C-1 and C-2 of the glycerolbackbone (for 18:0, 283)
M-$O(CH_2)_{15}CH_3$	159	Removal of the O-alkyl moiety (m/z, 187 for 18:0)
Acetyl ion	43	An expected release (same for 18:0)

In addition to examination of the diacetates, a particularly effective assay technique involves formation of the isopropylidene derivative as described in Figure 4-22. The product, which is obtained in 95–98% yield, is good proof for the presence of a 1-(3)-O-alkylglycerol since this acetonation reaction will occur only with vic-glycol type structures. Thus this would eliminate substitution of an ether bond at the sn-2 position. The isopropylidene derivative can be analyzed by GC and/or GC-MS; these techniques reveal excellent patterns, which are very valuable for proof of structure of the ether as such and for

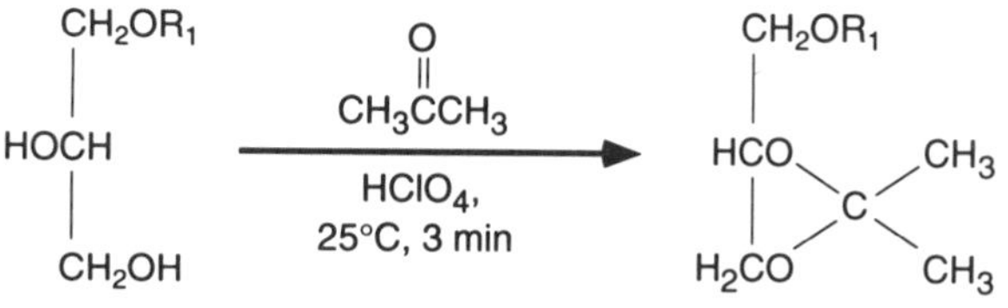

FIGURE 4-22. Conversion of 1-(3)-alkylglycerol to an isopropylidene derivative.

identification of the hydrocarbon side chain. In another way, the iso-propylidene derivative can be most helpful, and that is in the bulk preparation of specific chain lengths through the use of preparative gas-liquid chroma-tography. The isolated derivative can be converted to the free glyceryl ether form by treatment with dilute acid.

There has been an enormous amount of literature on the mass spectrometry of lipids; if any insight into this area is desired, it is suggested that reference be made to Egge (1983) and Murphy (1993).

An Additional Mass Spectrometric Approach   While the specific mass spectrometric analysis just described works well with the alkylglycerols, an equally informative route to the definition of the structure of the ether-linked phosphoglycerides is through examination of the FAB-MS spectrum of the alkyl (lyso)glycerophosphocholine. The latter derivative is prepared by base-catalyzed methanolysis as described earlier in this chapter. The lyso com-pound, purified by thin-layer chromatography, is mixed with thioglycerol and placed on the probe for the FAB-MS assay. A typical spectrum would reveal the following peaks: $[MH]^+$, protonated mass ion: 482 (for 16:0 chain length), 510 (for the 18:0 chain length) and 508 (for the 18:1 chain length);

$$m/z \ 224 \ \text{for} \ CH_2{=}CH{-}CH_2 O \overset{\overset{\displaystyle O}{\|}}{\underset{\underset{\displaystyle O}{\|}}{P}} OCH_2CH_2N(CH_3)_3 \ \text{and} \ m/z \ 184 \ \text{for the} \ O\text{-}$$

phosphocholine moiety. The data support the conclusion that this ether-linked phosphoglyceride is indeed an alkyl(ether)glycerophosphocholine. However, as stressed earlier, this methodology will not provide information on the stereo configuration of the molecule.

Periodate Oxidation.   As described earlier, periodic acid will cleave vici-nal hydroxyls in a quantitative manner. In the case of 1-*O*-alkylglycerol (and also 3-*O*-alkylglycerol), oxidation of 1 mol of glyceryl ether will yield 1 mol of formaldehyde, $H_2C{=}O$, and 1 mol of an alkoxyacetaldehyde, $ROCH_2CHO$. The reaction is quantitative at room temperature and at neutral pH.

Periodic acid action on a typical synthetic glycerylether (e.g., 1-*O*-hexadecylglycerol) will show the presence of 1 mol of *vic*-glycol per mole of starting substrate. This type of data is expected for a synthetic 1-*O*-

hexadecylglycerol, but use of periodic acid alone will not prove whether a 1-*O*- or 3-*O*- glyceryl ether (or a mixture) was present. It would, however, eliminate the presence of a 2-*O*-glyceryl ether, which would be unreactive to periodic acid. More refined analytical tools, as will be described below, are needed to establish the stereochemical configuration as well as other characteristics.

Thus, at this point, one can rest assured that a glyceryl ether is present in the presumed alkylacylGPC, and it is either the 1- or 3-*O* type. The latter question will be addressed in sections below.

STEREOCHEMICAL CONFIGURATION.   While the optical rotatory values obtained with highly purified alkylacylphosphatidylcholine, $[\alpha]_D^{25} + 2.5 - 3.0°$ are suggestive of an *sn*-3 configuration, this information alone cannot be considered as positive proof. The most effective route to determination of the stereochemical configuration of an ether-linked phosphoglyceride is through the use of phospholipase A_2. Using the enzyme purified from the venom of *Naja naja*, cleavage of the fatty acyl substituent on a naturally occurring alkylacylcholine phosphoglycerides can proceed very easily to completion. The experimental conditions are exactly the same as described earlier in this chapter for phospholipase A_2 attack on a diacylphosphatidylcholine. The fact that the phospholipase A_2 can remove the fatty acyl group from the alkyl ether-linked phosphoglyceride constitutes proof for the *sn*-3 configuration for this molecule. All saturated ether containing choline phosphoglycerides examined to date also possess the *sn*-3 stereochemical configuration. In the same enzymatic reaction conducted on the alkylacylphosphatidylcholine, the liberated fatty acid can be identified further by the methodologies described earlier in this chapter. The fatty acyl groups, as expected, are almost exclusively of the unsaturated type.

An additional structure proof assay involves the use of phospholipase C isolated from *Bacillus cereus*. This reaction, which is conducted in the exact same way as described earlier in this chapter, yields an alkylacyl glycerol and an *O*-phosphocholine. These two products can be analyzed by the same procedures as outlined at that point.

The formation of an alkylacylglycerol and *O*-phosphocholine will provide additional excellent proof in support of the chemical structure of the alkylacylphosphatidylcholine.

Taken together, the above information strongly supports the structural formula of naturally occurring alkylacylcholine phosphoglycerides to be 1-*O*-alkyl-2-*O*-acyl-*sn*-glycero-3-phosphocholine.

Other Analytical Techniques

Several other analytical procedures can be of value, to varying degrees, in the proof of structure of glyceryl ethers and the ether-linked phosphoglycerides. Among these are the old-fashioned approaches: melting points and elemental analyses, exploration of their optical activity, and spectroscopic examination including infrared spectroscopy. A description of these assay systems and their merits and concomitant demerits is given below.

MELTING POINT VALUES AND ELEMENTAL ANALYSES.   These two analytical techniques are useful in following the progress of a synthetic procedure leading to a specific glyceryl ether or an ether-linked phosphoglyceride. The discussion here will center on the glyceryl ethers, which have been derived by Vitride reduction of the parent phosphoglyceride. These ethers are converted to their isopropylidene derivatives and subjected to preparative gas-liquid chromatography on ethylene glycol succinate (Hanahan et al., 1963). The desired peaks are collected and subsequently cleaved to the free glyceryl ether by treatment with 2 N HCl. The saturated components can then be crystallized from *n*-hexane and obtained as shiny crystals. As examples, 1-*O*-octadecylglycerol(batyl alcohol) melts at 70.5–71.0°C and 1-*O*-hexadecylglycerol (chimyl alcohol) melts at 64.5–65.0°C. This type of analysis obviously cannot be done with the unsaturated glyceryl ethers, which are usually liquids at room temperature. Nonetheless, the technique does have merit in the specific situation mentioned above. It is a borderline diagnostic test.

Elemental analyses for carbon and hydrogen are easily accomplished on the saturated glyceryl ethers, which were obtained as crystals above. These assays can be conducted by any one of the numerous microanalytical organic chemical laboratories in the commercial sector. Unfortunately, the unsaturated glyceryl ethers, which are usually liquid at room temperature, pose a problem. However, these ethers can be dissolved in chloroform–methanol (1:1, v/v) and their total weight determined by vacuum evaporation at 40°C in a tared flask. A similar aliquot can then be subjected to the same analyses as for the saturated type. The carbon and hydrogen content of 1-*O*-octadecyl glycerol (batyl alcohol) and 1-*O*-hexadecyl glycerol (chimyl alcohol) is C, 73.24 and H, 12.79 and C, 73.14 and H, 12.6, respectively. While these values reflect only a very small percentage difference and could not be used as indicators of purity, they would be important in establishing the structure of a product obtained by synthetic procedures.

OPTICAL ACTIVITY.   Optical activity measurements on the glyceryl ethers are performed on the isopropylidene derivatives since the glyceryl ether per se has such a low optical rotation value, $[\alpha]_D^{25}$ +2.5°. Some typical values for the acetone derivatives are as follows:

Compound	$[\alpha]_D^{25}$
1-*O*-Hexadecylglycerol	−16.7°
3-*O*-Hexadecylglycerol	+16.0°
1-*O*-Octadecylglycerol	−16.2°
3-*O*-Octadecylglycerol	+16.0°
1-*O*-*cis*-9-Octadecenylglycerol	−15.3°

These data show clearly that one can distinguish very easily the 1-*O*-configuration from the 3-*O*- forms. Naturally occurring glyceryl ethers found in the phospholipids in bovine erythrocytes and *Tetrahymena pyriformis*, for example, all show $[\alpha]_D^{25}$ values in the range of $-16.0°$ to $-16.5°$, a range indicative of the 1-*O* form. As noted earlier, the alkyacylphosphatidylcholines are low rotators, $[\alpha]_D^{25}$ $+ 2.5°$ to $3.0°$, so the best route to establishing the structural configuration of the glyceryl ether moiety is through recovery of the free glyceryl ether as described above and then examining the optical behavior of the isopropylidene derivatives (comparing to synthetic standards, if possible).

INFRA-RED SPECTROMETRY.   This technique has been applied with some success in identifying the ether bond, C—O—C in the free glyceryl ether, by its absorption band at 1100 cm^{-1}. It is not possible to use this absorption band to detect the ether bond in the ether-linked phosphoglycerides since it is swamped out by the very intense phosphate absorption in the region from 1100 to 1000 cm^{-1}. Thus it is mandatory to isolate the glyceryl ether component per se, if this type of information is desired. At best, the use of infrared spectroscopy is of limited value in any structural sense for the glyceryl ethers.

Alkenylacylphosphatidyl choline (1-*O*-*cis*-Alkenyl-2-*O*-acyl-*sn*-glycero-3-phosphocholine): plasmenylcholine; plasmalogen

A Brief Historical Perspective

The presence of an unusual phospholipid in mammalian cells that would yield, under acid conditions, a substance giving a positive reaction with the fuchsin-sulfurous acid reagent was reported in the period from 1927 to 1930. It was suggested an aldehyde had been released during this acid treatment. Yet it was very difficult to obtain the product in an unaltered form. Ultimately in the 1950s it was proven conclusively that an ether linked phosphoglyceride with 1 mol of a postulated vinyl ether residue and 1 mol of a fatty acid ester per mole of lipid P existed in mammalian cells. At the present time, vinyl-ether-containing phosphoglycerides have been found primarily in the ethanolamine phosphoglyceride fractions with lesser amounts in the choline-rich phosphoglyceride fractions.

A very low level of a vinyl ether component has been reported in the serine phosphoglycerides, but not in any other types. Over the next 20 years, this field of study developed in a significant manner; the contributions of a number of capable investigators are summarized in a very clear manner in a review by Debuch and Seng (1972).

Unfortunately the current interest level in the vinyl ether phosphoglycerides (plasmalogens) has been on the decline. This is partly due to the inherent instability of these molecules, which causes many investigators to avoid them, and partly due to the lack of any particular role (or metabolic activity) of these molecules or any of their derivatives in biological systems.

General Qualitative and Quantitative Assay Techniques

There are several procedures that can provide some information on the chemical nature of an alkenylacylcholine phosphoglycerides. Again, at this point as with the alkyl ether acylcholine phosphoglycerides, it is assumed that a highly purified sample (containing no diacyl or alkylacyl components) is available. Later the methodology to be used with mixtures will be discussed. Certain of these maneuvers have been mentioned earlier in this and in the previous chapter.

QUALITATIVE IDENTIFICATION: THIN-LAYER CHROMATOGRAPHY. Even though it is not possible as yet to distinguish the alkenylacyl, alkylacyl, and diacyl species by their R_f values on thin-layer chromatography, there is one spray that will show whether an aldehydogenic compound is present. Basically this color test depends on the classic interaction of an aldehyde with the Schiff reagent which is composed essentially of a 4–5% fuchsin solution (*p*-rosaniline, a heterocyclic toluidine derivative with a free amino group) mixed with a 10% sodium bisulfite solution diluted 15-fold with water. After standing for 12 hr, the solution is treated with charcoal and filtered. The filtrate can be stored in a dark bottle for a reasonable period of time.

Then, subsequent to chromatography of a plasmalogen containing sample on silica gel G, 250 μm, in chloroform–methanol–water (65:35:7, v/v), the plate is sprayed with the Schiff reagent. If any free aldehyde is present, a vivid violet color will result within 15–20 sec. The background is usually a very delicate purple color. If no color develops, another comparable plate is run; but in this case the plate is sprayed first with the Schiff reagent, followed by a spray with 0.1 M mercuric chloride. The latter reagent will cause cleavage of any vinyl ether linkage present, liberating a free aldehyde. The latter then reacts with the Schiff reagent, yielding the beautiful violet color.

Another spray used to detect the vinyl ether phosphoglycerides on thin layer chromatograms is one containing 2,4-dinitrophenylhydrazine in 2N HCl or 2 N H_2SO_4. However, when using this spray, it is first necessary to run the sample, spotted on a precoated silica gel G plate, 250 μm, in chloroform–methanol–water (65:35:7, v/v). At the end of the development, the plate is allowed to air dry and then is exposed to HCl fumes (usually the plate is suspended above a dish containing 12 N (concentrated) HCl for several minutes). It is then sprayed with the dinitrophenylhydrazine reagent. If an aldehyde is present, a yellow spot will appear within a few minutes at most.

It is important to stress that these tests represent presumptive evidence for a vinyl ether containing phosphoglyceride (plasmalogen), but not a final proof of structure. Other tests that can be used in conjunction with the above qualitative detection systems are the phosphorus spray and sulfuric acid char reaction to support the association of an organic phosphate containing compound with a plasmalogen positive spot. Such information is of considerable value in exploring further the structure proof on vinyl ether-linked phospholipids.

BASIC QUANTITATIVE PROCEDURES.   Phosphorus, nitrogen, and choline assays were described earlier for the diacyl and the alkylacyl analogs: Analysis for phosphorus and choline should give a molar ratio of 1.0. A similar ratio should be obtained from phosphorus and nitrogen analyses. If a dry weight can be obtained on a sample, then the phosphorus content should be near 4%. Other methodologies are outlined as follows.

Iodination.   The vinyl ether content of a phosphoglyceride sample can be assessed quantitatively by the iodine uptake method of Siggia and Edsberg (1968). Vinyl ethers react faster than olefinic double bonds, and hence this differential reactivity forms the basis of the assay. The reaction proceeds as prescribed in Figure 4-23. This principle was used in a modification of the reaction by Gottfried and Rapport (1962). In essence they developed a spectrophotometric procedure which measured the change in absorbance (at 355 nm) as the iodine interacted with the vinyl ether. A sample containing as little as 0.03 μmol (20–25 μg total weight) of ether-linked phosphoglyceride could be detected with ease. The molar extinction coefficient of iodine was reported to be 27,500.

Dinitrophenylhydrazine Reaction.   This assay is based on a method developed by Wittenberg et al. (1956). It takes advantage of the interaction of *p*-dinitrophenylhydrazine with a free aldehyde to yield a yellow-colored hydrazone derivative. The absorbance of the reaction mixture can be read on a spectrophotometer at 395 nm, and aldehyde can be determined quantitatively by this procedure. The reader is urged to consult this very informative article.

These assays are useful in providing information on the general character of a vinyl ether containing phosphoglyceride and support for an alkenylacyl-phosphatidylcholine structure. However, as emphasized previously in this book (perhaps *ad nauseum*), these assays will not provide a complete structure proof assessment. A more sophisticated approach, similar to that described

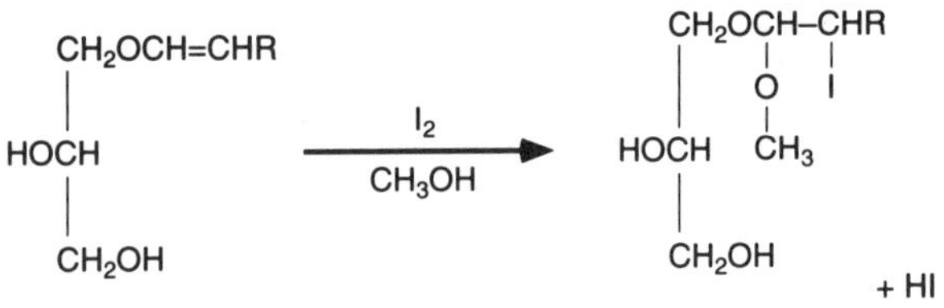

FIGURE 4-23. Iodination of an alkenylglycerol.

for the alkylacylphosphatidylcholine in this chapter, is needed. Some suggested routes to structure proof on the alkenyl ether acylphosphoglyceride are outlined as follows.

Specific Structure Proof Procedures Using Chemical
and Biochemical Reagents

The proof of structure of the vinyl ether-containing phosphoglycerides places some constraints on the methodology to be used. This is due to the inherent lability of the unsaturated ether linkage, $CH_2OCH=CH$, in the parent molecule. However, if proper care is exhibited in the handling and storage of these compounds, meaningful results can be obtained. The methods outlined here are by no means exhaustive, yet they can provide considerable insight into the structure of these unique phosphoglycerides.

BASE-CATALYZED METHANOLYSIS.   Using essentially the same procedures as described for proof of structure of the alkylacylphosphatidylcholine, as described earlier in this chapter, release of the fatty acyl substituent as a methyl ester can be achieved in a quantitative manner. At the same time, the monoalkenyletherphosphoglyceride (the lyso derivative) is produced. Only one word of caution: Neutralization of the alkaline reaction mixture must be done with great care since any excess acid will cause loss of the vinyl ether moiety. It is possible to use an ion exchange resin to neutralize the base. Experimentally the final reaction mixture is extracted as described earlier and subjected to thin layer chromatography. The methyl esters and the lyso-alkenyletherphosphoglycerides are isolated and separated as outlined above.

The methyl esters of the long-chain fatty acids can be subjected to exactly the same examination as outlined earlier in this chapter. In the usual case, these fatty acid esters will be composed largely of unsaturated (olefinic) linkages with little or no saturated components. This follows the pattern noted before for the substituents located at the C-2 position in other phosphoglycerides. A quantitative analysis of these fatty acids will show that there is 1 mol per mol of lipid phosphorus. The lysoalkenyletherphosphatidylcholine can then be studied further.

FAB-MS PROFILE.   This technique has been used to a limited extent to establish certain structural characteristics of this lysophosphoglyceride (Hanahan et al., 1990). Using the same procedure as discussed earlier in this chapter, the following major ions were noted for a highly purified (naturally occurring) sample: $[MH]^+$, molecular mass ions at m/z 480, m/z 510, and m/z 508. These ions were representative of vinyl ether homologs of side chain 16:0, 18:0, and 18:1, respectively. A base peak at m/z 184 was indicative of O-phosphocholine.

However, in any examination of the FAB-MS profile of these vinyl ether-linked compounds, one should be aware of some possible artifact formation depending on which matrix is used, thioglycerol or glycerol. In a study designed to explore the structure of an analogous derivative (i.e., 1-*O*-alkenyl-*sn*-glycero-3-phosphoethanolamine), Weintraub et al. (1991) compared the FAB-MS profile in a glycerol as well as a thioglycerol matrix. The fragmentation pattern in each case was similar, except that a cluster of ions, m/z 331, m/z 357, and m/z 359 was present in the sample run in thioglycerol, but was not present in the sample run in a glycerol matrix. The nature of this cluster suggested that the alkenyl side chains were associated with these fragments. Subsequent experimentation suggested that in the slightly acid environment of the thioglycerol, fatty aldehydes could be liberated from the alkenyl(lyso)-GPE and interact with the thioglycerol to yield the following derivative:

$$CH_2\!-\!CH\!-\!CH\!-\!S\!-\!CH\!=\!CHR_1$$
$$\quad\;|\qquad\;|$$
$$\quad OH\quad OH$$

Alternatively, there could be a direct attack by the thioglycerol on the vinyl ether bond. Subsequent experiments supported the formation of the derivative noted earlier. A similar reaction was found to occur with alkenylglycerophospho-choline.

Even though there was evident interaction of the thioglycerol matrix with the vinyl ether phosphoglyceride (in a very reproducible manner), this fact should not deter one from using this method for analysis of these compounds. FAB-MS provides excellent insights into the structure of this type of phospholipid.

CLEAVAGE OF THE VINYL ETHER BOND.   It had been shown many years ago that the vinyl ether bond was very sensitive to acid treatment, wherein a free long-chain aldehyde is formed. The mechanism of this cleavage is considered to occur first by protonation of the unsaturated linkage to form a carbonium ion. The latter then reacts quite rapidly with water to yield a free fatty aldehyde and an alcohol. This reaction works well with the lysoalkenyletherphosphatidylcholine and has the advantage that there are no interfering fatty acid esters.

Then one must be prepared to analyze this aldehyde quickly since it can be oxidized quite easily. A suggested assay system is one described by Mangold and Totani (1983), in which these aldehydes are subjected to gas liquid chromatography on 10% Silar 5CP on Gas-ChromQ (80–100 mesh) at 220°C. Suitable standards should be run and, of course, a mass spectrometer can be interfaced. A complete quantitative analysis of the fatty aldehydes can be achieved in this way.

The sensitivity of the vinyl ether bond to acid formed the basis of a method for determination of the vinyl ether-linked phosphoglycerides in a mixture of naturally occurring diacyl, alkyacyl, and alkenylacylphosphoglycerides (Murphy et al., 1993). Such mixtures were treated with acid and then subjected to high-performance liquid chromatography. The presence of a lyso(monoacyl)phosphoglyceride was taken as an indicator of the amount of vinyl ether-linked components present in the mixture.

An additional approach to identification of the vinylether side chain is through formation of the dimethyl acetals,

$$R—C \begin{matrix} OCH_3 \\ | \\ —OCH_3 \\ | \\ H \end{matrix}$$

Acid treatment of of the alkenyl(lyso)glycerophosphocholine in the presence of methanol can lead to quantitative formation of dimethyl acetals. The latter can be purified on silica gel G thin layer chromatography plates using a solvent system of petroleum ether (30–60°)–diethyl ether–acetic acid, (80:20:1, v/v). The acetals can be located with the TNS spray reagent. The positive areas migrating at R_f 0.65–0.68 are removed by scraping, and the acetals are extracted with chloroform–methanol–water (1:2:0.8, v/v). GC-MS analysis of the dimethylacetals can be accomplished as described by Hanahan et al. (1990). A standard dimethyl acetal can be prepared from a commercially available palmitaldehyde (hexadecylaldehyde) bisulfite complex.

HYDROGENOLYSIS.   This methodology involves the reaction of Vitride (sodium bis(2-methoxyethoxy)aluminum hydride with the lysoalkenyletherphosphatidylcholine as outlined for the saturated ethers earlier in this chapter. This leads to the formation of 1-O-alkenylglycerol in very good yields. This latter material can be recovered from the reaction mixture by solvent extraction with diethyl ether. The isolated alkenylglycerol can be subjected to the following reactions.

Periodate Oxidation.   It is possible to employ the same procedure as was used to prove the presence of *vic*-glycol groups in saturated glyceryl ethers. The important aspect of this assay is to maintain the pH of the reaction mixture at 7.0.

As described by Albro and Dittmer (1968), periodate oxidation of the alkenylglycerols proceeded smoothly to completion in a buffered mixture within 3 hr at room temperature. Basing calculations on the amount of lipid phosphorus used in the hydrogenolysis experiment just discussed, 1 mol of the 1-O-alkenylglycerol will yield 1 mol of formaldehyde. As little as 0.1 μmol of alkenylglycerol can be determined.

This assay provides excellent support for the presence of a vinyl ether on either the *sn*-1 or *sn*-3 position of a glycerol backbone.

Diacetates.   The alkenylglycerols are isolated by thin-layer chromatography on silica gel G in a solvent system of hexane–diethyl ether (80:20, v/v). These compounds can be located by use of the TNS spray or by running an additional thin-layer chromatography plate for charring with sulfuric acid.

The desired areas are removed from the TNS-sprayed plate by scraping and then extraction with diethyl ether–hexane (1:10, v/v).

Subsequently the alkenylglycerols can be reacted with acetic acid anhydride–pyridine, usually a 1:5 (v/v) mixture, in a sealed tube. The tube is heated at 70–80°C for 45 min and then cooled to room temperature, and the seal is carefully broken. The contents are diluted with an equal volume of water and extracted with *n*-hexane. Two separate extractions with *n*-hexane should allow complete recovery of the alkenylglycerol diacetates. The combined hexane extracts are washed with water until neutral and then dried over anhydrous Na_2SO_4. The purity of the preparation can be determined by thin-layer chromatography on silica gel G in a solvent system of petroleum ether–diethyl ether–acetic acid (80:20:1, v/v). Again two separate plates can be run, one for spraying with TNS and the other for charring with sulfonic acid (plus heat).

Characterization of these diacetates can be accomplished by use of gas-liquid chromatography. A reversed-phase column (e.g., 3% SE 30) can be employed at 240°C with helium as the flow gas. The 16:0 chain length will elute in approximately 10 min, followed by the 18:0 and 18:1 chain lengths in the range between 15–20 min. Interfacing with a mass spectrometer will allow acquisition of the EI mass spectrum for further identification of the structural features of the molecules.

An adjunct to these methods is to place the diacetates on a thioglycerol matrix and examine its fragmentation pattern by FAB-MS, as described previously. In this instance, though, one should be aware of the artifactual cluster of ions at m/z 331, m/z 357, and m/z 359, which are the result of interaction of thioglycerol with the vinyl ether bond. Nonetheless, FAB-MS is a potent accessory in these structure proof studies.

CONFIGURATIONAL ANALYSIS.   Two characteristics of the choline-containing vinyl ether phosphoglyceride remain to be established, namely, the geometric configuration of the vinyl ether bond (whether *cis* or *trans*) and the stereochemical conformation of the glycerol backbone (whether *sn*-1 or *sn*-3). In the latter case, an *sn*-2 conformation is excluded since the periodate oxidation results suggest either an *sn*-1 or an *sn*-3 form. The experimental approach to resolution of the (above) two questions follows.

Geometric Configuration.   The establishment of the chemical nature of the ether linkage in the "plasmalogens" as a vinyl ether immediately raised the question of its geometric configuration (i.e., *cis* or *trans*). The answer to this question rested solely on the use of the infra-red spectral properties of naturally occurring alkenyl ether derivatives with standard synthetic vinyl ethers. Norton et al. (1962) used a synthetic 1-butenyl ether as a standard, whereas Warner and Lands (1963) employed the *cis* and *trans* forms of synthetic methyl 1-dodec-1-enyl ether as a standard for comparison.

In the latter study, Warner and Lands (1963) prepared alkenylglycerol through use of phospholipase C on the parent alkenylacylglycerophosphocholine. Subsequent base-catalyzed methanolysis of alkenylacylglycerol yielded alkenylglycerol, which could be converted to alkenylglycolaldehyde by periodate oxidation. Thus the only difference between the standard synthetic alkenyl ether and the alkenylglycolaldehyde was the carbonyl function in the latter derivative:

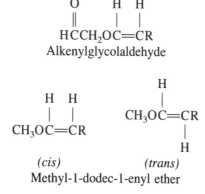

$$\begin{array}{c}O \quad\;\; H \;\; H \\ \| \qquad | \;\; | \\ HCCH_2OC{=}CR\end{array}$$

Alkenylglycolaldehyde

$$\begin{array}{cc}H \;\; H & H \\ | \;\; | & | \\ CH_3OC{=}CR & CH_3OC{=}CR \\ & | \\ & H\end{array}$$

(cis)   *(trans)*
Methyl-1-dodec-1-enyl ether

The synthetic *cis* standard and the alkenylglycolaldehyde showed identical infrared spectra, with comparable peaks at 1670, 1270, 1110, and 735 cm^{-1}, except for a peak at 1750 cm^{-1}, attributable to the carbonyl function. Absorption peaks at 1200 cm^{-1} and 930 cm^{-1}, representative of a *trans*-alkenyl ether, were not present. In their study, Norton et al. (1962) also presented evidence for the occurrence of a *cis* configuration for the vinyl ether bond in naturally occurring material. It is important to note that examination of the infrared spectrum of the intact alkenylacylglycerophosphocholine was not possible for identification of *cis* or *trans* ether bonds. This was due to the fact that the phosphate esters and the unsaturated fatty acids absorb strongly at wavelengths critical to distinguishing the *cis* and *trans* alkenyl ethers. Finally, Cymerman-Craig and Hamon (1965), provided further confirmatory evidence that the vinyl ether bond in naturally occurring alkenylether phosphoglycerides was of the *cis* configuration.

Absolute Configuration.   The correct stereochemical configuration of naturally occurring alkenylacylglycerophosphocholine could not be obtained by direct comparison with synthetic derivatives since the latter were simply not available. However, two other routes to definition of the stereochemical configuration were available, one a physical-chemical approach and the other a biochemical method.

Using a physical-chemical protocol, Cymerman-Craig et al. (1966) employed phospholipase C attack on the naturally occurring alkenylacylglycerophosphocholine to yield alkenylacylglycerol. The latter derivative was subjected to base-catalyzed methanolysis to produce alkenylglycerol. The latter compounds were converted in part to the diacetates and in part hydrogenated and then converted to the diacetates. The optical rotatory dispersion pattern of these derivatives in ethanol were examined down to 232 nm. The data showed very clearly that the alkenylglycerols and their hydrogenated derivatives, the alkylglycerols, had the same absolute configuration as naturally occurring 1-*O*-hexadecylglycerol (chimyl alcohol) and 1-*O*-octadecylglycerol (batyl alcohol). These results supported an *sn*-1 configuration for the naturally occurring alkenylethers.

In the biochemical method, the enzyme phospholipase A$_2$, isolated from *Naja naja* snake venom can attack the native alkenylacylglycerophosphocholine and liberate completely the esterified fatty acid and the alkenyl(lyso)glycerophosphocholine. On the basis of the stereospecific mode of attack of this enzyme on the 2-acyl ester position of *sn*-3 phosphoglycerides, it can be concluded that the naturally occurring alkenylacylglycerophosphocholine possessed the *sn*-3 stereochemical configuration.

COMMENTS ON STRUCTURE PROOF STUDIES.   Considerable emphasis has been placed on methodology that can be used in studies designed to prove the structure of the various forms of phosphatidylcholine. By no means has it been an exhautive survey, but it was designed to provide basic insights into the process of identification and characterization of the complex lipids. Also each class of compound described here was assumed to be pure—that is, free of any other classes. Certainly, as shall be described next, this is not the case in the usual biological sample, but again a "scientific license" was invoked by the author to illustrate the chemistry of these compounds. Furthermore, it was assumed in the aforementioned descriptive material that sufficient amounts of material were available to undertake all of the analytical procedures. This, of course, is not the always the case when one has only a limited number of cells available for such an adventure. In this case, one has to choose the methodology that best suits the goals of the experiment. In any event, caution should be exerted in extending a phosphorus value and an R_f value to a full structure proof decision. Cowardice can form the best part of valor at this point.

A Reality Check: Cellular Diacyl, Alkenylacyl, and Alkylacyl, Glycerophosphocholines and Their Detection

In the real world, many cells contain not just one class but rather three classes of choline-containing phosphoglycerides—that is, diacylglycerophosphocholine, alkenylacylglycerophosphocholine, and alkylacylglycerophosphocholine. Thus, in any study designed to investigate the potential involvement of these cellular phosphoglycerides in a signal transduction process, it is important to prove whether all of these lipids are involved or whether only one particular class is important. Even if only one class appears to be associated with a particular agonist's interaction with a cell, a further inspection may reveal that only one particular species within that class of phosphoglyceride is implicated. An example of the latter phenomenon would be the specific release of the long-chain fatty acid, arachidonic acid. Thus, it is important to be able to identify and differentiate, in particular, these classes of phosphoglycerides. The following discussion centers on the methodological approach to this problem.

Two different procedures can be used to determine the distribution of the above three classes of phosphoglycerides in a cellular lipid preparation. The first involves an enzymatic cleavage and the second involes a chemical approach. These are outlined briefly as follows.

Enzymatic Attack

The lipids isolated from a cell preparation are separated on thin-layer chromatography, and the desired fractions are recovered by the extraction procedure described earlier in this chapter. In this instance, interest is centered on the phosphatidylcholine fraction, and it can be subjected to the protocol outlined in Scheme A (Figure 4-24).

Solvent system 1 is composed of petroleum ether (30–60°C)–diethyl ether–acetic acid (90:10:1, v/v) and solvent system 2 is toluene. Using this

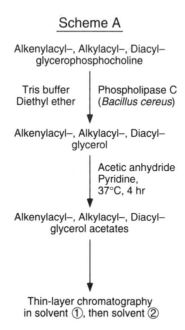

Scheme A

Alkenylacyl–, Alkylacyl–, Diacyl–
glycerophosphocholine

Tris buffer | Phospholipase C
Diethyl ether | (*Bacillus cereus*)

Alkenylacyl–, Alkylacyl–, Diacyl–
glycerol

                | Acetic anhydride
                | Pyridine,
                | 37°C, 4 hr

Alkenylacyl–, Alkylacyl–, Diacyl–
glycerol acetates

Thin-layer chromatography
in solvent ①, then solvent ②

FIGURE 4-24. An enzymatic approach to the cleavage of a mixture of diacyl-, alkylacyl-, and alkenylacyl-GPC to the corresponding "diglycerides" by use of phospholipase C. The diglycerides are then converted to their acetate derivatives (Scheme A).

protocol, Sugiura et al. (1983) showed that the three classes of lipid acetates were well separated from each other. The alkenylacylglycerol acetates migrated the fastest and the diacylglycerol acetates moved the slowest. The alkylacylglycerol acetates migrated to an R_f value intermediate between the other two types. These derivatives can be located by TNS spray and extracted, and their composition can be determined by methods outlined earlier. The extent of the analyses depends on the goals of the experiment and the amounts of lipid available.

Chemical Cleavage

This protocol involves first a base-catalyzed methanolysis and then an acid-mediated reaction as outlined in Scheme B (Figure 4-25).

A particularly good analytical tool to use in this experimental protocol is the phosphorus assay. Thus, if the total P content is known, the percent distribution in the various fractions can be obtained. For example, the P value of fraction I will relate back to the diacylglycerophosphocholine level. Then analysis of fraction II will allow calculation of the alkenylacylglycerophosphocholine value. Finally the P content of fraction III will reflect the al-

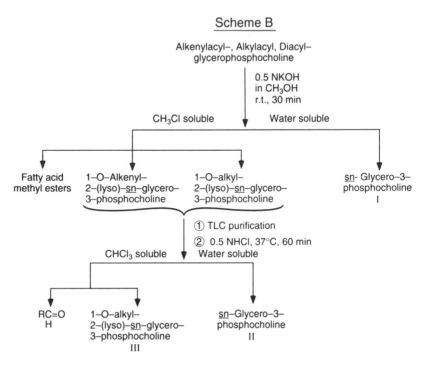

FIGURE 4-25. A chemical approach to analysis of a mixture of diacyl-, alkylacyl-, alkenyl-GPC (Scheme B).

kylacylglycerophosphocholine level. Other analyses can be conducted depending on the objectives of the experiment and the total amount of sample available for use.

A Final Group of Choline-Containing Phospholipids: The Sphingophospholipids

The only other major choline-containing phospholipid found in mammalian cells is sphingomyelin. Usually it is present in relatively small amounts (~5–8% of the total phospholipid), and until rather recently it was considered to be mainly a structural component of a membrane. However, there is growing evidence supporting a possible role of sphingosine-containing phospholipids in signal transduction processes. The latter topic will be addressed briefly as follows.

A common feature of these phospholipids is that they contain a long-chain aminodiol, sphingosine, as the basic backbone of the molecule. In mammalian cells this long-chain base is the predominant form with much smaller amounts of a saturated derivative, dihydrosphingosine. One should be aware of the possible presence of long-chain bases, other than sphingosine, or dihydrosphingosine in the sphingomyelin isolated from a mammalian cell or system (Karlsson, 1970). In this section, however, it is assumed, for illustrative purposes, that sphingosine is the only long-chain base found in the sphingomyelin sample (Figure 4-26).

As noted, until fairly recently, sphingomyelin has been thought to be a relatively inert (metabolically) molecule in cells, with its main feature being that of a structural component. However, current studies in several different laboratories suggest that metabolites, such as sphingosine-1-phosphate, sphingosine, and sphingosine phosphocholine, are involved in the signal transduction pathway, especially as regulators of phospholipase C.

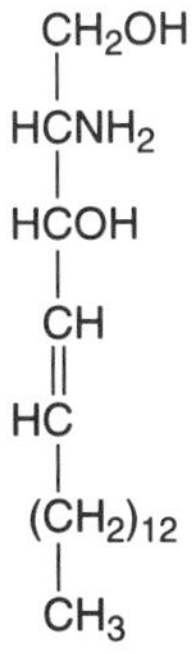

CH_2OH
|
$HCNH_2$
|
$HCOH$
|
CH
||
HC
|
$(CH_2)_{12}$
|
CH_3

Sphingosine

FIGURE 4-26. Chemical structure of naturally occurring sphingosine.

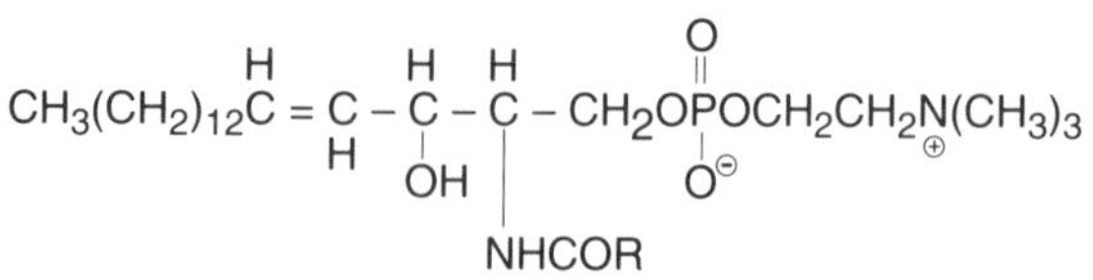

$$CH_3(CH_2)_{12}\overset{H}{\underset{H}{C}}=\overset{H}{C}-\overset{H}{\underset{OH}{C}}-\overset{H}{C}-CH_2O\overset{O}{\underset{O^{\ominus}}{P}}OCH_2CH_2\overset{\oplus}{N}(CH_3)_3$$

NHCOR

Sphingomyelin
(*N*-acyl-sphingosylphosphocholine)

FIGURE 4-27. Chemical structure of naturally occurring sphingomyelin.

Some Comments on Nomenclature and Structure

Even though it is commonly referred to as sphingosine, its proper chemical name is D-*erythro*-1,3-dihydroxy-4,5-*trans*-octadecene. Other satisfactory names are *trans*-4-sphingosine, sphing-4-enine, or sphingenine; in any event, all the naturally occurring sphingolipids contain only the D-*erythro* form.

In normal mammalian cells, free sphingosine, as such, is not found or is present in extremely low amounts. Usually sphingosine is found as a more complicated structure with a long-chain fatty acyl group in amide linkage with the C-2 amino function and a polar head group, usually O-phosphocholine, associated in ester linkage to the C-1 hydroxy function. The resulting compound, sphingomyelin—or, using chemical nomenclature, *N*-acylsphingosyl-phosphocholine—can be illustrated by the conventional formulation shown in Figure 4-27. While this structural formula is a convenient one for general use, in actual fact a more accurate representation for this molecule (in an aqueous environment) would be one in which the long-chain hydrocarbon of the fatty acyl substituent and the long-chain hydrocarbon of the sphingosine backbone are arranged in a parallel configuration as shown in Figure 4-28. This latter presentation has much in common with that of the diacylphospho-glycerides, which can be shown in a comparable configuration. Physical chemical measurements—for example, surface behavior and X-ray spectra of several sphingosine derivatives—support such a physical arrangement.

Isolation and Characterization

While sphingomyelin is found in most mammalian cells, it is likely to represent only 5–10% of the total phospholipid. However, there are, as expected, exceptions to this postulate. In the human erythrocyte sphingomyelin repre-

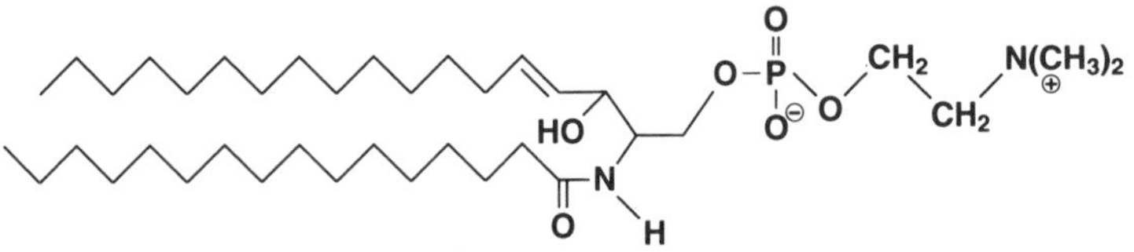

FIGURE 4-28. Another chemical depiction of sphingomyeline based on its orientation in an aqueous medium.

sents, as expected, 5–10% of the total phospholipid, but in the bovine erythrocyte, it amounts to nearly 50% of the total phospholipid. Little or no phosphatidylcholine is present in the latter cell.

Typical Isolation Procedure

In chapter 3 the experimental route to isolation of individual classes of phospholipids from cellular preparations was described in some detail. Either a column-chromatographic or a thin-layer chromatographic (TLC) procedure can be used here. If preparative TLC plates (normally silica gel G) are available, the separation of sphingomyelin from its most likely contaminant, phosphatidylcholine, is easily accomplished. The only other probable contaminant would be monoacylglycerophosphocholine (lysolecithin), but it is usually present in very, very low concentrations. If only a relatively few cells are available for lipid extraction, the TLC route is the procedure of choice. If milligram quantities of sphingomyelin are desired, then the cell of choice is the bovine erythrocyte and the isolation can be accomplished as described by Hanahan (1961).

Characterization

Several analytical techniques can be employed to characterize and prove the structure of sphingomyelin isolated from cells.

PHOSPHORUS, NITROGEN AND CHOLINE CONTENT.   Assuming sufficient material has been isolated so that a total weight value can be obtained, then exactly the same methodology as described in Chapter 3 can be used. Thus, the phosphorus and nitrogen assays can be accomplished easily by standard digestion procedures. The N/P molar ratio of a pure sphingomyelin sample should be 2.0. The release of choline for assay will, however, require more stringent conditions than encountered with the phosphoglycerides. In a typical hydrolytic procedure, the sphingomyelin sample is reacted with 2N HCl in 80% aqueous methanol for 16 hr at 80–85°C. The reaction flask is cooled, and the contents are mixed with an equal volume of petroleum ether (30–60°C). The petroleum ether-rich upper phase contains the methyl esters of the amide linked fatty acids. This fraction can then be saved for further analysis.

The water–methanol-rich lower phase is treated with an alkaline solution to convert the sphingosine "acid salt" to a neutral form. This mixture is extracted with diethyl ether to recover the free sphingosine, which can then be analyzed further. The water-rich phase from the latter extraction can be analyzed for choline as described in Chapter 3. The choline/P molar ratio should be 1.0. An alternative route to the release of O-phosphocholine and choline by action of phospholipases C and D, respectively, will be considered below.

FATTY ACID ASSAY.   In the procedure just outlined for the degradation of sphingomyelin, one of the products is the methyl ester of long-chain fatty acids. These are easily checked for purity by thin-layer chromatography using silica gel G-coated plates (250 μm) in a solvent system of petroleum ether (30–60°)–diethyl ether–glacial acetic acid (80:20:1, v/v). The esters will migrate to an R_f value near 0.85 and can be extracted with petroleum ether (30–60°)–diethyl ether (80:20, v/v). The extract is then analyzed by gas-liquid chromatography or the latter combined with a mass spectrometric procedure.

These fatty acid esters will show a composition distinctly different from that of a phosphoglyceride, such as diacylphosphatidylcholine. The first difference is that there usually are little or no unsaturated fatty acids present in the sphingomyelin-derived material. The second difference is that the saturated fatty acids derived from sphingomyelin, isolated from bovine erythrocytes, contain a high percentage (46%) of 22:0 and 24:0 chain lengths. The remaining 54% are composed of 16:0 and 18:0 chain lengths. Often when sphingomyelin, isolated from bovine erythrocytes, is chromatographed on a silica gel G-coated plate (250 μm) in a solvent system of chloroform-methanol–water (65:35:6), two spots can be detected. This is attributable to separation of the 16:0/18:0 pairs from the 22:0/24:0 pairs.

This fatty acid composition results in a more rigidly parallel structure, which has led to the assumption that these lipids are only structural components of biological membranes and not metabolically active (or activatable) compounds. This assumption may have to be revised pending continuing studies on the role of sphingosine-like compounds in signal transduction.

SPHINGOSINE.   The sphingosine formed in the hydrolytic cleavage of sphingomyelin, as described above, can be characterized by two procedures. The most effective one is to acetylate it with acetic anhydride in pyridine, which will interact with the two hydroxyl functions and the free amine group. The product is reacted with alkali in methanol, which will release the acetyl groups from C-1 and C-3 of the sphingosine but not from the amine (amide) function. This product is then treated with a trimethylsilylating agent (which reacts with the two free hydroxyls), yielding a derivative that can be analyzed by gas-liquid chromatography–mass spectrometry. Usually a column containing 3% SE 30 or OV-1 will provide a good chromatographic profile of the sample (Carter and Gaver, 1967). Sphingosine can be analyzed also through initial formation of an O-phthalaldehyde derivative followed by examination of its HPLC profile on a reversed-phase (C_{18}) column (Wilson et al., 1988). The level of detection is in the picomole range. Pure sphingosine is available from commercial suppliers and can be used as a control. Another analytical procedure would be to subject the sphingosine sample to periodate oxidation. The products would be 1 mol of a long-chain aldehyde, 1 mol of formaldehyde, 1 mol of formic acid, and 1 mol of ammonia from 1 mol of sphingosine.

Infrared Spectrocopy and FAB-MS

The infrared spectral pattern of sphingomyelin has limited use in establishing the structure of this compound. However, there are a few absorption bands of supportive value in characterizing this molecule. A strong absorption band at 1650^{-1} cm supports the presence of amide linkage. The presence of a *trans* double bond produces a significant absorption band at 975 (980–970) cm^{-1}. The absence of a band at 1750 cm^{-1} would show that the sample was free of an ester carbonyl function, such as found in phosphatidylcholines and the glycoglycerolipids.

Fast atom bombardment–mass spectrometry has been used to a limited extent in characterizing sphingomyelin. In a thioglycerol matrix, the following major ions were detected: MH$^+$, 731; M-28, 703; m/z 239, $CH_3(CH_2)_{12}CH{=}CH{-}CH^+$; m/z 184, O-phosphocholine (Dr. S. Weintraub, personal communication).

Location of Phosphate Ester

Both phospholipase C and phospholipase D can attack sphingomyelin to give N-acylsphingosine plus O-phosphocholine and N-acylsphingosylphosphoric acid plus free phosphoric acid, respectively. These enzymatic reactions are conducted in exactly the same way as described earlier in this chapter for attack on the phosphatidylcholines. A sphingomyelin-specific phospholipase C is available from commercial suppliers as are other phospholipase C preparations with broader specificity.

The action of phospholipases C and D certainly provided proof for the attachment of the phosphate group as an ester and also provided proof that the choline was linked covalently to the phosphate. However, their action did not prove where the phosphate was located, nor did it prove whether phosphocholine moiety was attached to the C-1 or C-3 hydroxyl function. Basic phosphorus analysis on a weighed sample of sphingomyelin proved that there was only 1 mol of phosphorus per mole of sphingomyelin. Proof of the location of the O-phosphocholine was afforded by a strictly chemical approach. In Figure 4-29 the sphingomyelin is reacted with formic acid and hydrogen peroxide to give the trihydroxy derivative. The latter was then subjected to cleavage by periodate, followed by permanganate oxidation and finally acid hydrolysis. Serine was recovered in good yield and proved that the O-phosphocholine must have been attached to the C-1 hydroxyl in sphingomyelin. These reactions are outlined in Figure 4-29.

A Short Diversion: The Sphingoglycolipids

While the main thrust of this book remains centered on the phosphorus-containing lipids of mammalian cell membrane, it would be remiss on the author's part not to discuss the sphingoglycolipids. These phosphorus-free, sphingosine-containing lipids coexist with the phospholipids in cellular membranes. In place of a phosphocholine moiety (as in sphingomyelin), a carbohydrate is substituted. These complex lipids were found to be present in significant amounts in brain some 60 years ago, and their potential association with

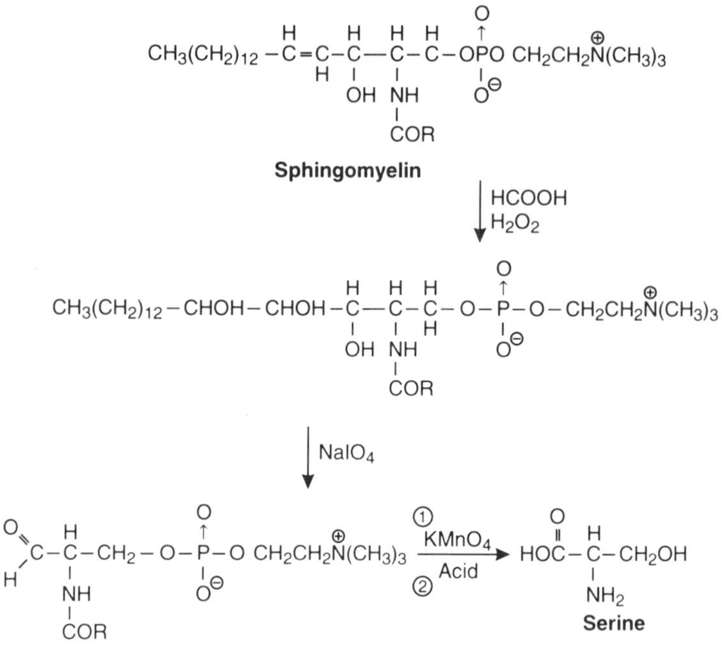

FIGURE 4-29. A chemical approach to proof to structure of sphingomyelin.

certain disease states prompted a tremendous interest in their chemical structure and in their biochemical behavior in the ensuing years. The basic details of these important studies are provided by Kanfer and Hakomori (1983). A few of the more salient features of the chemical nature of these compounds will be discussed here.

Basically there are two main classes of these compounds, and their general characteristics are explored in the next section.

Cerebrosides

These sphingolipids are composed of fatty acids (in amide linkage), a long-chain base (most commonly sphingosine), and a carbohydrate (usually galactose, sometimes glucose) in a glycosidic linkage. The structure of a galactocerebroside is given in Figure 4-30.

The structure proof of the naturally occurring cerebrosides was obtained by several different maneuvers. First, since the cerebroside was nonreducing, the carbohydrate must be linked through the aldehyde group. Second, methylation of cerebroside, followed by hydrolysis, yielded 2,3,4,6-tetramethylgalactose. In addition to supporting a glycosidic bond between the sphingosine and the C-1 position of the galactose, evidence is provided for the 1,5-pyranose structure of this carbohydrate. Further confirmation of the attach-

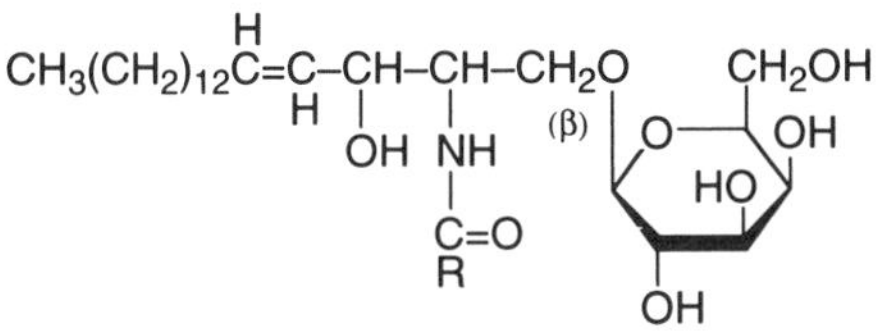

$$CH_3(CH_2)_{12}C=C-CH-CH-CH_2O$$

FIGURE 4-30. The chemical structure of a naturally occurring galactocerebroside.

ment of the carbohydrate to the C-1 hydroxyl group on sphingosine was obtained by acetylation of the cerebroside, followed by hydrogenolysis over a platinum catalyst. In this reaction the allylic acetoxy group on C-3 is released and the double bond is hydrogenated. A subsequent hydrolytic procedure yielded sphingine—that is, 2-amino-octadecanol-1. On the basis of a comparable reaction product with standards of well-defined structures, the conclusion was reached that the galactose was attached to the C-1 position of sphingosine.

In addition to the cerebrosides, which contain only one carbohydrate residue, there are other glycosphingolipids in mammalian cells that contain more than one sugar component. These oligosaccharide derivatives are called globosides. For example, lactosyl ceramide (1-O-lactosyl-N-acylsphingosine) is a constituent of the erythrocyte membrane. Ceramide trihexoside accumulates in the kidneys of patients with Fabry's disease, due to the lack of a lysosomal α-galactosidase A activity.

Cerebroside Sulfuric Esters (Sulfatides)

Approximately 20% of the cerebrosides in the white matter of brain is in the form of sulfuric acid esters. These compounds are also found in general in nervous tissue. It was long suspected that the sulfate group was attached to the galactose residue, but the exact location was not established. Current evidence supports the attachment of the sulfate group to the C-3 position of the galactose moiety. This decision was reached as follows: Brain sulfatide was hydrolyzed and the liberated sugar sulfate was isolated and permethylated. Subsequent reaction with strong acid in the presence of aniline yielded the anilide of 2,4,5,6-tetramethylgalactose. Since the hydroxyl function of the original sugar sulfate was not methylated, it was evident that the sulfuric acid must have been esterified to the C-3 hydroxyl in the original sulfatide.

Gangliosides

The second major class of sphingoglycolipids found in mammalian cells is the ganglioside. It was first isolated from the brains of humans who died as the result of Niemann-Pick disease and later was also identified in those persons who were diagnosed with Tay-Sachs disorder. The gangliosides are present in normal brain tissue, particularly localized in the ganglion cells and have been found in other cells—for example, the erythrocyte and the spleen.

These complex glycolipids contain N-acylsphingosine, to which is attached (in glycosidic linkage) lactose (glucosylgalactose), N-acetylneuraminic acid (a monoamino, monobasic polyoxy acid with nine carbon atoms), N-acetyl-galactosamine, and finally a terminal galactose molecule. This compound is called monosialoganglioside I. Other variations of this structure have also been found in naturally occurring sources. These molecules, as might be expected, are quite water-soluble, and this characteristic forms the basis for their recovery from cells.

Isolation and Detection

Though a detailed examination of this subject is beyond the scope of this book, a few general comments are in order. A favorite isolation procedure entails use of the Folch-Pi–Lees–Sloane-Stanley partition procedure (1957), as modified by Suzuki (1965), which involves use of chloroform, methanol, and water containing 0.88% KCl. Two phases are formed in this partition method, and the water-rich upper phase contains over 90% of the polar gly-colipids, including the gangliosides. Phospholipids are well separated into the chloroform-rich phase. However, if a significant amount of calcium ion is present, this will complicate the isolation procedure. In the latter case the gangliosides will be found in the chloroform-rich layer. Thus it is important to keep the calcium level very low so that nearly quantitative recovery of the gangliosides in the water-rich phase can be achieved. The details of the isolation, detection, and structure proof of cerebrosides, cerebroside sulfate, and gangliosides is masterfully outlined by Hakomori (1983). Thin-layer chromatography is most helpful in following the progress of the isolation of these complex lipids. In a neutral solvent system (chloroform–methanol–water (65:35:6, v/v) and using a silica gel G-coated plate (250 μm), the cerebrosides will migrate to an R_f value of 0.70, the sulfatides will migrate to an R_f value of 0.40, and the gangliosides are found at R_f 0.25. Two-dimensional thin-layer chromatography is particularly useful in the separation of complex mixtures of these glycolipids.

Certain spray reagents can be helpful in locating these carbohydrate-containing sphingolipids. Resorcinol yields a blue violet color with gangliosides, and naphthol produces a blue color with cerebrosides, cerebroside sulfuric acids, and gangliosides.

Of importance the sulfatides can be further distinguished by a more sophisticated procedure. This would entail intraperitoneal injection of $^{35}SO_4$ into small animals. A quite rapid incorporation of label into brain lipids, for example, will provide support for the presence of a sulfatide. The isolated sulfatide-containing fraction is subjected to thin-layer chromatography, and the lanes are sectioned and assayed for radioactivity. Location of labeled material migrating at the same R_f value of standards (nonlabeled) would be supportive for the occurrence of a sulfatide. If further examination of the product is desired, the reader is urged to consult a chapter by Kanfer (1983). A word of caution must be given at this point since there is very good evidence for the occurrence of the sulfate moiety in the glycoglycerolipids. For further detailed information on this subject, consult the chapter by Murray and Narasimhan (1990).

Notes on Analogous Compounds: The Glycoglycerolipids

Another group of carbohydrate-containing lipids, other than the sphingoglycolipids, have been found in many tissues and are referred to as glycoglycerolipids. Instead of a sphingosine backbone, these compounds utilize glycerol. A typical example would be galactosyldiacylglycerol {1,2-di-*O*-acyl[β-D-galactopyranosyl(1→3)]-*sn*-glycerol}, whose structure is shown in Figure 4-31.

There are several other, more complicated forms of the glycoglycerolipids, which contain alkylacyl groups in place of the diacyl residues plus the addition of sulfate and other carbohydrate groups. In addition to the galactoglycerolipids, several comparable (structurally) glucoglycerolipids have been discovered.

Another Summation and Another Interface

This concludes a discussion of the choline-containing phosphoglycerides and the sphingosine-containing phospholipids. The next chapter will focus on another set of phosphoglycerides important as membrane components, namely, the ethanolamine-, inositol-, and serine-containing phosphoglycerides. While it may seem that there has been an overemphasis on the chemistry of the choline phosphoglycerides and the sphingophospholipids, the information provided here can be applied to other phosphoglycerides and phospholipids as well. As stated before, this book was not intended to be a *tour de force* in lipid chemistry, but rather a modest introduction to the chemical nature and the chemical identification of the phospholipids. In the ensuing chapter, reference will be made to techniques outlined in this and earlier chapters. It is hoped

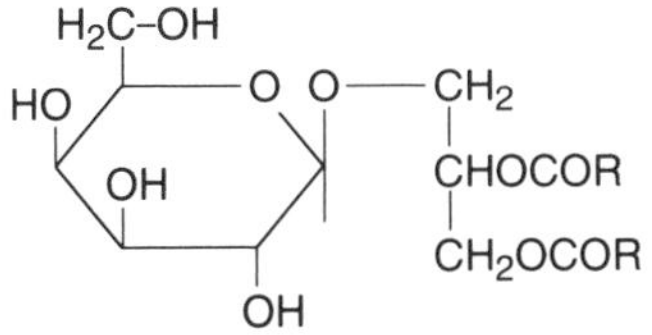

FIGURE 4-31. The chemical structure of a typical glycoglycerolipid, in this case galactosyldiacylglycerol.

that the importance of positive proof for the chemical structure of any phospholipid implicated in a biological reaction or system is a self-evident truth. In support of the latter statement, a relatively simple, nearly quantitative chemical procedure for the structure proof of platelet activating factor and related compounds has been developed and is described in Chapter 6. This technique holds the promise that it will be applicable to other phosphoglycerides discussed in this and the following chapter.

FIVE

NON-CHOLINE-
CONTAINING
PHOSPHOLIPIDS

Diacyl-, Alkylacyl-, Alkenylacyl-
Ethanolaminephosphoglycerides,
Phosphatidylinositols, and
Phosphatidylserine

A Continuing Saga

The diversity of phospholipids present in the mammalian cell membranes continues to titillate one's scientific senses. In addition to the presence of at least 12 different structural types of phospholipids in a cell that serve to complicate the picture, there are species within species. If one considers the diacyl, alkylacyl, and alkenylacyl variants, plus the number of different fatty acyl and fatty ether combinations, there can be several hundred different species present. Certainly progress is being made in relating certain species with a particular cellular process, and this is no doubt an exciting and important area of study. However, this is only the tip of the "cellular iceberg," since there is little or no information on the biological role of the majority (certainly over 75%) of the phospholipids. Questions to be asked center on the need for such a spectrum of phospholipids. Are some structural components only, are some vestigial remnants, or do they play a crucial role in biological reactions yet to be discovered? There is no simple answer as yet, but this trend of thought should be kept in mind in any investigation on membrane lipid behavior. An important route to interpreting the role of various phospholipids in a biological milieu is to be certain of the chemical structure and identification of the molecules under study. So in continuation of the general format used in Chapter 4, the chemistry of the ethanolamine-, inositol-, and serine-containing phosphoglycerides will be explored at this point. A limited excursion will be made as to their participation in biological reactions.

Though the above three classes of compounds share certain common structural features, there are sufficient differences to warrant separate treatment of each group of compounds. For example, the ethanolamine-containing phosphoglycerides can contain, in addition to the diacyl form, an alkylacyl and/or alkenylacyl form. Inositol-containing phosphoglycerides other than the diacyl type have not been reported, but several other phosphorylated species have been detected. The serine-containing phosphoglycerides have been found only as diacyl derivatives.

A Common Physical-Chemical Feature of Non-Choline Containing Phosphoglycerides

Often one will see these phosphoglycerides referred to as "acidic" phospholipids. Certainly the chemical nature of the substituent attached to the phosphoric acid residue—for example,

$$O-\overset{\overset{\displaystyle O}{\|}}{\underset{\underset{\displaystyle O^-}{|}}{P}}\text{-ethanolamine}, \quad O-\overset{\overset{\displaystyle O}{\|}}{\underset{\underset{\displaystyle O^-}{|}}{P}}-O \text{ inositol},$$

and

$$O-\overset{\overset{\displaystyle O}{\|}}{\underset{\underset{\displaystyle O^-}{|}}{P}}-O \text{ serine}$$—would support such a classification. The charge char-

acteristics of a series of phosphoglycerides, at various pH values, were established in a particularly able manner by Garvin and Karnovsky (1956). In this study the titration of phosphatidylserine and phosphatidylethanolamine (as well as phosphatidic acid and phosphatidylcholine) in 2-ethoxyethanol was investigated. While phosphatidylinositol (or any of its phosphorylated derivatives) were not examined (only phosphatidylinositol had been shown to be a membrane lipid at that time!), one can extrapolate the results obtained with the other compounds to the phosphoinositides. Essentially the following charge profile was noted:

Phosphatidylethanolamine is a zwitterion over the pH range of 2–7; and in the range of 7–10, it is in the anionic form.

Phosphatidylserine, as expected, exhibited a more complex titration pattern. At pH 7.0, the diester phosphoric acid and carboxyl groups were in the anionic form, whereas the α-amino group had a positive charge. Below pH 4.5, phosphatidylserine is in the zwitterion form. Interestingly, this compound, as isolated from several different tissues, usually contains 1 mol of a monovalent cation per mole of phosphoglyceride.

Phosphatidylcholine is a zwitterion over the entire pH range, as shown by the lack of any titratable groups from pH 1 to 12.

The ionic characteristics of phosphatidylethanolamine and phosphatidylserine provided a basis for their separation from the neutral phospholipids, such as phosphatidylcholine and sphingomyelin. This involved application in chloroform–methanol (1:1, v/v) of a cellular lipid extract to a neutral aluminum oxide column (0.5 mg phospholipid P per gram of adsorbent) in the same solvent mixture. The zwitterionic phosphoglycerides (i.e., the choline-containing types) are not retained by this adsorbent and quickly pass through the column. The acidic phospholipids (i.e., phosphatidylethanolamine, phosphatidylserine, phosphatidylinositol, and its phosphorylated forms) phosphatidic acids, and others are strongly adsorbed to this column in this solvent. These phospholipids can be eluted in bulk by passing several volumes of ethanol–chloroform–water (5:2:2, v/v) through the column. This procedure illustrates quite clearly the decided charge characteristics of certain phosphoglycerides and consequently allows a simple straightforward method for complete separation of the choline-containing from the non-choline-containing phospholipids. This approach has merit in that further fractionation of the various species can be more easily accomplished. When a complex lipid extract is chromatographed on a single adsorbent (e.g., silicic acid), there tends to be significant overlap of some of the species due to loading factors and the long elution time required for the elution process. Also, the chance in the latter case for degradative changes is much greater.

Structural and Other Facets of the Non-Choline Containing Phosphoglycerides

Cellular Distribution

The levels of these lipids vary from cell to cell. In general, though, the ethanolamine phosphoglycerides are the predominant species, ranging from 22% to 29% of the total phospholipids. The phosphoinositides are found in much lower abundance: Phosphatidylinositol is 2–7% of the total phospholipids, and their more complex phosphorylated forms one much less than 1%. Interestingly, phosphatidylserine constitutes approximately 12–19% of the total phospholipid phosphorus. For comparison purposes, the choline phosphoglycerides are close to 35–45% of the total cell phospholipid in most cells. Exceptions to the latter are the bovine and sheep erythrocytes, where little or no choline-containing phosphoglycerides are found. In the latter cells, sphingomyelin is the only choline containing phospholipid at levels of 45–55%.

Phosphatidylethanolamine

In common with choline-containing phosphoglycerides, the ethanolamine-containing phosphoglyerides can be found in three different chemical forms— that is, diacyl-, alkylacylacyl-, and alkenylacylglycerophosphoethanolamine. If these three types are present in a cellular lipid extract, it is not possible to separate them cleanly by column chromatography or thin-layer chromatography. Differentiation of these species must be accomplished by a chemical means similar to that described for their counterparts in the choline phosphoglycerides (see Chapter 4). At this point, the approach will be to use diacyl-glycerophosphoethanolamine (diacylphosphatidylethanolamine) as a model compound and illustrate how its chemical characteristics and structure can be determined. To a large extent, then, this methodology will be applicable to alkylacyl and alkenylacyl forms mentioned above. A discussion of the assay of a mixture of these three forms will be considered later.

Isolation and Purification

As described earlier, the total cellular lipids can be recovered by use of a neutral organic solvent system, such as chloroform–methanol–water. Silicic acid column chromatography, thin-layer chromatography, and high-pressure liquid chromatography (HPLC) are well suited to isolation of the ethanolamine-rich phospholipids. Only column and thin-layer chromatographic purification will be discussed at this juncture.

SILICIC ACID COLUMN CHROMATOGRAPHY.   This procedure is very effective in the preparation of milligram quantities of phosphatidylethanolamine. An effective protocol would involve use of an adsorbent, such as Silic-AR CC-7, dispersed in chloroform and loaded into a glass column measuring 4 cm × 20 cm. In a typical separation, approximately 20 mg of lipid P, in chloroform, is carefully loaded onto 40 g of silicic acid previously packed into the above column. Sequential elution is then initiated as follows: chloroform, 100 ml; chloroform–methanol (20:1, v/v), 150 ml; chloroform–methanol (6:1, v/v), 500 ml; methanol, 200 ml. If necessary to improve flow of solvent, pressure can be applied in the form of (compressed) nitrogen gas. Incidentally, columns with as little as 5 g of silicic acid can be used with success (with a similar lipid P/adsorbent ratio of 1:2). In any event, pure phosphatidylethanolamine is found in the chloroform–methanol (6:1, v/v) eluate. This fraction is evaporated to dryness in a rotary evaporator, and the residue is dissolved to volume in chloroform–methanol (2:1, v/v) and stored at −20°C in the dark until needed. This will be referred to below as the total phosphatidylethanolamine fraction and is usually recovered in a 90% yield.

THIN-LAYER CHROMATOGRAPHY.   If desired, the phosphatidylethanolamine (PE) component can be isolated from the total cellular lipid using 500- or 1000-μm precoated silica gel G plates Loading ratios for the 500-μm plates (20 × 20 cm) can be as high as 400 μg of lipid P. Solvent systems can be *acidic* (chloroform–methanol–acetic acid–water (60:35:1:8, v/v), *neutral* (chloroform–methanol (56:35:6, v/v), or basic (chloroform–methanol–28% ammonium hydroxide, 70:30:5, v/v). In the neutral and acidic solvent systems the R_f value for PE will be close to 0.60, while in the basic system the R_f will be near 0.45. Using the detection reagents cited below, the desired area can be located. It is removed by scraping, and the silica gel is extracted with chloroform–methanol–water (1:2:0;8, v/v). Usually two separate extractions with this solvent will allow a good recovery (over 80%). Phasing by the addition of chloroform and water will distribute the PE into the chloroform layer. The latter can be evaporated to dryness under nitrogen, dissolved in chloroform–methanol (1:1, v/v), and stored at −20°C until needed.

Two more recent routes to the detection and assay of ethanolamine and related compounds deserve mention here. Sundler and Akesson (1975) reported an elegant method for the analysis of ethanolamine and possible derivatives (e.g., O-phosphoethanolamine) using the dansyl(5-dimethylaminonaphthalene-1-sulfonyl) derivatives and their fluorescence characteristics, in the 0.05–5 μM range. McMasters and Choy (1992) outlined a procedure for the determination of ethanolamine, by reverse-phase HPLC, as the phenythiocarbamyl derivative. A quantitative evaluation could be obtained in the 0.1–10 nmol range.

Detection and Assay Procedures

The PE sample obtained by column chromatography or thin-layer chromatography can be subjected to the following techniques.

DETECTION WITH SPRAY REAGENTS.   In exactly the same manner as described in Chapter 3, use of the phosphorus and ninhydrin sprays plus the char reaction will provide a reasonable qualitative assessment of the purity of the sample.

ANALYTICAL THIN-LAYER CHROMATOGRAPHY.   This methodology is particularly helpful in establishing the apparent purity of a PE sample. This is the case with very small amounts of lipid (nanograms) encountered in signal transduction experiments on cells. If a radioactive compound is used in the experiments, then a standard, nonlabeled, PE sample can be chromatographed on an adjacent lane on the thin-layer plate. This lane can be sprayed with the P and ninhydrin reagents to locate the PE, and then the area adjacent to this area on the (experimental) lane can be marked, extracted, and analyzed for radioactivity and lipid P, assuming that sufficient material is present.

The solvent systems described immediately above are useful in evaluating the general character of the PE sample. Only one word of caution, as regards use of the basic solvent systems, is that the plate must be thoroughly air-dried before using the ninhydrin spray.

In the PE fraction obtained from certain tissues or cells, there may be apparent "contaminants" revealed on thin-layer chromatography as slower moving components. These P positive spots may represent the presence of the monomethyethanolamine and/or the dimethyethanolamine analogs. An additional methyl substituent would yield the quaternary derivative, or the choline analog.

PHOSPHORUS, NITROGEN, AND ETHANOLAMINE ASSAYS.   Aliquots of the PE sample can be analyzed quantitatively for total phosphorus and nitrogen using the techniques described in Chapter 3. The molar ratio of phosphorus to nitrogen should be 1.0 for diacylphosphatidylethanolamine. The ethanolamine assay will be described next.

Ethanolamine must be liberated from the PE sample by hydrolysis with acidified methanol for analysis. Subsequent to recovery of the water-soluble components, ethanolamine can be estimated by periodate oxidation and/or the dinitrofluorobenzene (DNFB) reaction (for colorimetric assay) as described by Kates (1972). However, these methods will not prove that the compound under study is ethanolamine.

If sufficient material is present, the dinitrophenyl derivative can be formed by reaction with DNFB in alkaline solution. The DNP-ethanolamine derivative can be isolated in the crystalline form, which melts sharply at 90°C. Elemental analysis of this derivative will provide proof of structure. Another approach will be to establish the presence of ethanolamine in the original PE sample by FAB-MS as will be shown as follows.

COMMENTS ON ANALYTICAL TECHNIQUES IN SIGNAL TRANSDUCTION EXPERIMENTS.   If one is investigating the behavior of the ethanolamine phosphoglycerides (or any other of the cellular lipids) in signal transduction experiments, then the central problem centers on the very small amounts of lipid material available in the usual experimental system. This difficulty can be circumvented through the use of radioactive (or radiolabeled) precursors such as [^{3}H]ethanolamine, ^{32}P$_i$, or ^{3}H-long-chain fatty acids (i.e., palmitic acid or arachidonic acid). In a typical experiment a cellular preparation (e.g., human platelets) is incubated with a radiolabeled precursor for 1–2 hr at 37°C, the platelets are washed and centrifuged, and the lipids are extracted in the usual way. This procedure was detailed in Chapter 4. This exercise centers on thin-layer chromatographic separation of the lipids in either a neutral or basic solvent system, using a synthetic or a highly purified naturally occurring phosphatidylethanolamine in an adjacent lane. In the case of tritiated precursors, the plate can be scanned on a BioScan System 200 Imaging Scanner (BioScan, Washington, D.C.) and the area(s) of radioactivity located. Then the lane, containing standard phosphatidylethanolamine, can be sprayed (covering the other lanes with a plain glass plate). If the area of the standard compares favorably (R_f-wise) with a radioactive area on the experimental lane, one can assume with quite reasonable certainty that the latter represents a radiolabeled ethanolamine derivative. Then, the silica gel is removed by scraping and is extracted with chloroform–methanol–water (1:2:0.8, v/v). Subsequent addition of chloroform and water to this extract will phase the labeled phosphatidylethanolamine into the chloroform-rich layer. The level of radioactivity in the sample can be determined by liquid scintillation counting and would represent a control value.

In a signal transduction experiment, human platelets are labeled as described above, a sample is removed as a control, the remainder is incubated with an agonist under specific conditions (closely regulated time, cell counts, and concentration), and the lipids are extracted and analyzed as described immediately above. Any significant changes in the level of radioactivity in the phosphatidylethanolamine fraction would suggest a metabolic alteration or turnover and an involvement in the signal transduction process.

However, the above observation will not provide any clue as to the mechanism, whether [³H]ethanolamine was lost by the action of phospholipase D (giving unlabeled phosphatidic acid and free labeled ethanolamine) or by that of phospholipase C (yielding a diglyceride and O-phosphoethanolamine). On the other hand, phospholipase A_2 activity would yield radiolabeled lysophosphatidyethanolamine which would migrate well below the starting material. In two of the above scenarios, detection of diglyceride and phosphatidic acid would not be feasible since they would be unlabeled. Two alternative possibilities present themselves. First, if there is a sufficient number of platelets, then there may be an adequate amount of material on the thin-layer chromatogram to detect with spray reagents (TNS, for example). The second intervention would employ another label. Incubation with radioactive phosphorus or tritiated palmitic acid would aid in identifying other products such as phosphatidic acid or diglyceride.

The above protocol may seem convoluted and mysterious, and it is; but there are not many choices available other than a detailed analytical approach as outlined earlier. Otherwise, a clear and meaningful interpretation of the role of phosphoglycerides in agonist-induced cell activation cannot be achieved.

Proof of Structure

The methodology used here for structure proof of phosphatidylethanolamine (and its analogs) is very similar to that described in Chapter 4 for phosphatidylcholine (and its analogs). Consequently a more limited discussion of the procedures and protocols for phosphatidylethanolamine will be undertaken here.

BASE-CATALYZED METHANOLYSIS. In the same way as it proved useful for structure proof with the choline-containing phosphoglycerides, base-catalyzed methanolysis is a most satisfactory approach to defining the structure of phosphatidylethanolamine. In this instance, it is expected that this reaction will yield methyl esters of long-chain fatty acids and glycerophosphoethanolamine. A brief description of the technique follows.

As before, the phosphatidylethanolamine sample in chloroform–methanol (1:10, v/v) is treated with 0.5 N NaOH in methanol for 20 min at room temperature. The reaction is stopped by the addition of an equivalent amount of 6 N HCl. Subsequent extraction of the mixture by the Bligh-Dyer method will ultimately yield a chloroform soluble fraction (containing the methyl esters) and a water-soluble phase (containing the glycerophosphoethanolamine). Under these conditions, the reaction is usually quantitative; this can be checked by assaying the water-soluble phase for total phosphate. Then the two fractions are analyzed as follows:

Methyl Esters.   These long-chain acyl esters are subjected to thin-layer chromatography, colorimetric analysis, and gas-liquid chromatography, coupled with mass spectrometry, as described in Chapter 4. These assays will provide information on the ester/P molar ratio (based on starting phosphate value it should be 2.0) and on the composition and relative distribution of fatty acyl residues.

Glycerophosphoethanolamine.   The water-soluble fraction is assayed for total phosphate (and as a control for inorganic phosphate). The latter is a check on whether the reaction proceeded as expected. It can be purified by ion exchange chromatography and then examined as follows.

Optical Activity.   Glycerophosphoethanolamine derived from a naturally occurring diacylphosphatidylethanolamine will show the following optical activity value, $[\alpha]_D^{25}$ −3.2° to −3.4°, as compared to an optical activity value, $[\alpha]_D^{25}$ −2.8° to −3.0°, for a synthetic *sn*-glycero-3-phosphoethanolamine. While these low optical activity values will not prove unequivocally the stereochemical configuration of the naturally occurring material, it is suggestive of an *sn*-3 configuration.

Periodate Cleavage.   Using a neutral periodate solution, 1 mol of glycerophosphoethanolamine will consume 1 mol of this reagent, with the liberation of 1 mol of formaldehyde. These results will prove the presence of a *vic*-glycol group in the starting material. It will not denote the stereochemical configuration of the sample.

It is possible to use a commercially available synthetic glycerophosphoethanolamine for comparison with that isolated from a natural source. In addition, there are synthetic and highly purified naturally occurring phosphatidylethanolamines available from commercial supply houses. This makes life easier for an investigator probing the structure of cellular lipids, but these compounds are low rotators and caution must be used in defining stereochemical configuration based solely on optical activity values.

STEREOCHEMICAL CONFIGURATION.   As with the diacylphosphatidylcholines, phospholipases can be used in assigning a configuration to the ethanolamine phosphoglycerides. Applying the same strategies as outlined in Chapter 4, phospholipase A₂ from cobra *(Naja naja)* venom or rattlesnake *(Crotalus adamanteus)* venom, is most effective in defining the stereochemical configuration of the ethanolamine phosphoglycerides. In addition, phospholipases C and D can provide excellent supporting proof of structure and configuration (phospholipase C). An expansion of this approach is presented as follows.

Phospholipase A₂ Action.   As in the case of phosphatidylcholine, the above-mentioned phospholipases will attack only the *sn*-3 form of naturally occurring (as well as synthetic) phosphatidylethanolamine. The products are, of course, lysophosphatidylethanolamine (1-*O*-acyl-2-lyso-*sn*-glycero-3-phosphoethanolamine) and the fatty acids (liberated from the *sn*-2 position). The latter can be analyzed for composition and structure, as the methyl esters, by gas–liquid chromatography coupled with mass spectrometry. Usually these acyl groups are largely the unsaturated types.

Assuming that the above enzymatic reaction was run in an ether-rich medium, the products can be isolated by thin-layer chromatography. Thus, the fatty acids released from the *sn*-2 position can be easily recovered as well as the lysophosphatidylethanolamine. The latter derivative will contain the fatty acyl groups associated with the *sn*-1 ester position. Base-catalyzed methanolysis of the lyso compound will produce the methyl esters. In the usual instance, these will contain mainly saturated chains. In any event the attack by phospholipase A₂ can proceed smoothly to completion. These results would strongly support an *sn*-3 stereochemical configuration for the parent diacylphosphatidylethanolamine.

Phospholipase C Attack.   As noted before, phospholipase C has limited stereochemical specificity, though it does attack an *sn*-3 phosphoglyceride at a considerably faster rate than the *sn*-1 configuration. One important value of this enzyme in structure proof studies lies in the fact that it can show quite conclusively that a diacylglycerol is attached via phosphate ester linkage to *O*-phosphoethanolamine. The reaction conditions and procedures are similar to those described in Chapter 4 for phospholipase C action on phosphatidylcholine. In the case of diacylphosphatidylethanolamine, the phospholipase C-derived (from *Bacillus cereus*) *O*-phosphoethanolamine can be analyzed by the procedure described by Sundler and Akesson (1975). In this case, the phosphoethanolamine-containing fraction is analyzed for total P and then treated with acid phosphatase to liberate the free ethanolamine. The latter can be analyzed spectrophotometrically as its DNS (5-dimethylamino-naphthyl-1-sulfonyl) derivative. This assay is linear in the range from 0.05 to 5 µM.

Phospholipase D Behavior.   While this enzyme shows no stereospecificity in its attack on the ethanolamine phosphoglycerides, its action in releasing free ethanolamine and a phosphatidic acid provides very clear evidence that the ethanolamine was attached to the parent molecule via a phosphate ester bond. Again, the same protocols and enzyme sources illustrated in Chapter 4 for phosphatidylcholine can be applied very easily to the ethanolamine phosphoglycerides.

A Short Summation.   The above experimental exercises provide evidence for an *sn*-3 configuration for naturally occurring phosphatidylethanolamine.

In concert with the other analytical data provided above, the structure of this type of phospholipid can be formulated as 1,2-diacyl-*sn*-glycero-3-phosphoethanolamine.

FAB-MS PROFILE.   While the FAB-MS is a powerful tool for certain facets of the identification of polar lipids, it does not lend itself easily to quantitation or to definition of the stereochemical conformation of a phospholipid. Nonetheless, it does provide information of supporting value for structure proof studies. An important difference between the choline-containing and the ethanolamine-containing phosphoglycerides is the occurrence of an m/z 142 in the spectrum of the latter compound in thioglycerol. It represents the protonated *O*-phosphoethanolamine.

BACK TO NEAR-REALITY: DIACYL-, ALKENYLACYL-, AND ALKYLACYLPHOS-PHOETHANOLAMINES.   Though in certain mammalian cells diacylphosphatidylethanolamine is the only ethanolamine phosphoglyceride present (e.g., liver), one is more likely to encounter in other cells a mixture of the diacyl alkenylacyl- and alkylacylethanolamine phosphoglycerides. In some cells, the alkenylacyl type accounts for nearly 70% of the fraction. The detection, analysis, and structure proof can take exactly the same form as described for the analogous choline derivatives. Another unique form of PE, the *N*-acyl derivative, has been found in mammalian cells and this general area has been reviewed by Schmid et al. (1990). Interestingly, *N*-arachidonoyl ethanolamine, which appears to be formed *in vivo* by phospholipase D action on the parent *N*-acyl PE, behaves like an endogenous cannabinoid (Devane et al. 1992). This fascinating area of research is the subject of intensive scientific research at present.

Phosphatidylinositols

If one, in response to some emotional stimulus, was moved to wax romantic about any cellular phospholipids, certainly the all encompassing adjective "glamorous" would be an apt choice for the phosphatidylinositols. Mention the magic phrases "signal transduction and phosphoglyceride turnover" and the sensitive, swift minds of many scientists will be programmed immediately to think of the inositol-containing phosphoglycerides. An even pedantic examination of the literature over the past 10–15 years will attest to the enormous attraction of these fascinating lipids to scientists of many different persuasions. When the additional magic word, second messenger, was used to describe the action of products of their metabolism in signal transduction systems, their place in the sun was firmly established and will not be easily overshadowed by any other phospholipid. There is much excellent scientific support for the importance of the phosphatidylinositols and their metabolic products in cellular processes. However, one must not lose sight of the fact that there are many other biologically active phospholipids in nature, which are waiting in the wings to be labeled "nearly glamorous."

In spite of the "tongue-in-cheek" remarks made above, there is no doubt that the phosphoinositides represent important components of the signal transduction pathway in many cells. As such, the results of investigations of these phospholipids have changed their role as structural components only to a more dignified position as metabolic "activist" in cells.

The specific impact of these phosphoglycerides on biological processes has developed from the fact that their hydrolysis during agonist interaction yields biologically active products. There is formation of the inositol phosphates, which are considered a second messenger(s), together with the diacylglycerols intimately associated with the activation of phosphokinase C. A less well publicized but equally important product can be arachidonic acid, which can be converted to biologically active prostaglandins.

As is to be expected, the literature on the biochemistry of the phosphatidylinositols in cellular reactions has been immense. No attempt will be made to cover this topic. Instead, only the salient features of the biochemical behavior of these compounds will be summarized in a concluding segment of this discussion. References to particularly informative review articles will be cited at that time. The main theme here, as in previous sections, will be devoted to the chemical characteristics and characterization of these most interesting phosphoglycerides.

In order to place the following discussion in proper perspective, it is worthwhile first to present the chemical structure of the three major inositol-containing phosphoglycerides found in mammalian cells. Their structural formulae are provided in Figure 5-1. Ptd Ins refers to phosphatidyl inositol, which also can be described at 1,2-diacyl-sn-glycero-3-phospho-1-D-myo-inositol. Of course the term Ptd (phosphatidyl) is most appropriate for indicating a diacylglycerophosphate residue. Ins obviously refers to the inositol residue. Ptd Ins (4)P then is simply the Ptd Ins molecule with an additional phosphate esterified to the 4 position of the inositol. Finally, Ptd Ins (4,5)P$_2$ signifies a Ptd Ins molecule with two phosphates esterified at the 4 and 5 positions on the inositol. Briefly the stereochemistry of myoinositol is such that the hydroxyl function at its 2 position is in an axial configuration whereas the rest of the hydroxyl functions are equatorial. In contrast to the phosphatidylcholine and phosphatidylethanolamine fraction in many cells, only the diacyl form is found with no evidence for the occurrence of any alkylacyl or alkenylacyl residues. These inositol-containing phosphoglycerides constitute 2–8% of the total phospholipids in mammalian cells, with 80–90% comprised of Ptd Ins.

As in a musical composition, there is a major chord, and on most occasions there is a minor chord. The same can be applied to phosphatidylinositol and its phosphorylated derivatives. Immediately above, there was mention of three major inositol-containing phosphoglycerides. Over the past few years, identification of so-called minor inositol phosphoglycerides has received in-

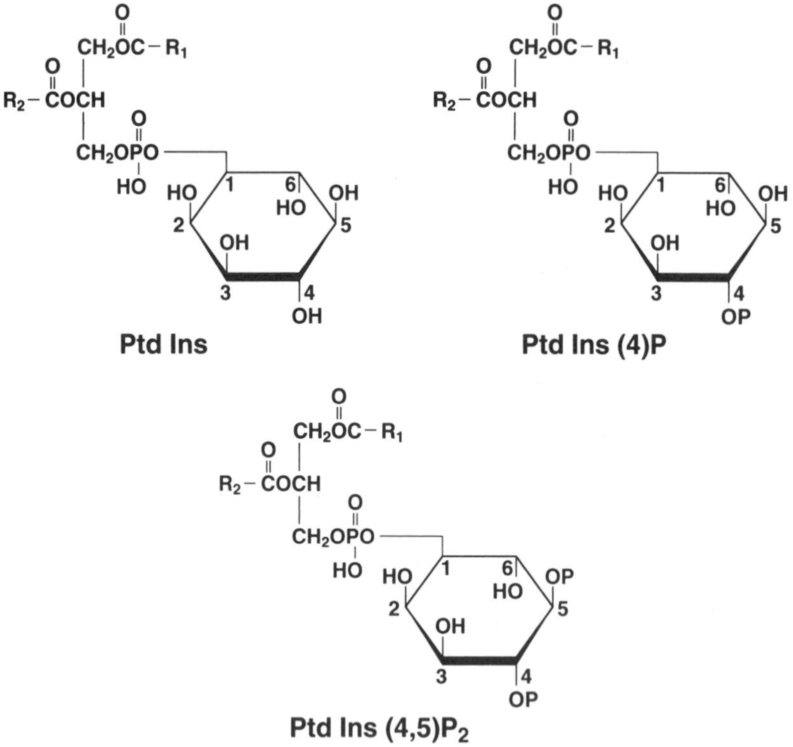

FIGURE 5-1. Chemical structure of three different, naturally occurring, inositol-containing phosphoglycerides.

creasing attention. Of particular interest, these new derivatives show a different phosphorylation pattern compared to three major types mentioned above. Examples of the new types are Ptd Ins (3)P, Ptd Ins (3,4)P_2 and Ptd Ins (3,4,5)P_3. Although these latter compounds are at most 2% of the total phosphatidylinositols, there is support for the fact that they may be more metabolically active than the "glamorous" major inositol phosphoglycerides. A fine discussion of these new derivatives can be found in a review by Stephens et al. (1993).

A Modest Historical Review

THE CHEMICAL SIDE.   The first evidence for the presence of inositol in naturally occurring phospholipids was reported by Folch (1942). He proposed that the inositol present in brain phospholipids was a meta-diphosphate (the nomenclature in vogue at that time). Some 7 years later, Folch (1949) provided data supporting a diphosphoinositide structure, composed of equimolar amounts of fatty acids, glycerol, and inositol metadiphosphate. In addition, this phospholipid as isolated could contain cations such as Ca^{2+}, Mg^{2+}, K^+, or N^+. In the Folch study, acid hydrolysis was used to liberate the inositol phosphate, but this treatment leads to an optically inactive *myo*-inositol phosphate. However, Pizer and Ballou (1959) established that base hydrolysis of a soybean phosphoinositide yielded an optically active *myo*-inositol phosphate, identified as L-*myo*-inositol-1-phosphate. These and other results suggested that the position of the inositol phosphate linkage in the parent molecule was similar. Grado and Ballou (1960) reinvestigated Folch's original findings and concluded that the acid hydrolysis of bovine brain phosphoinositides yielded optically active *myo*-inositol, mono-, di-, and possibly triphosphates. However, using base hydrolysis, Grado and Ballou obtained one *myo*-inositol, two *myo*-inositol diphosphates, and two *myo*-inositol triphosphates. All of the latter phosphates were optically active.

In independent observations, Dittmer and Dawson (1961), Dawson and Dittmer (1961), and Grado and Ballou, (1960) provided excellent data in support of the existence of an inositol-containing phospholipid of brain that yielded inositol triphosphate upon base hydrolysis. In addition, the presence of a monophosphoinositide, which on base hydrolysis yielded inositol mono-phosphate, was established with certainty. Further studies in several lab-oratories have supported the occurrence of three major inositolphosphates, namely, inositol-1-phosphate, inositol-1,4-diphosphate, and inositol-1,4,5-triphosphate as constituents of mammalian phosphatidylinositols. These phos-phates are linked via the 1-phosphate group to diacylglycerol in a D-config-uration. Hence a proper systematic name for a phosphatidylinositol would be 1,2-diacyl-*sn*-glycero-3-phospho-1-D-*myo*-inositol. The occurrence of an inositol-containing phosphoglycerides of the L-series has not been reported.

An excellent, well-written review (really a tome) on the stereochemistry of the inositol phospholipids is one by Parthasarathy and Eisenberg (1986). It should be on the "must read" list of investigators laboring in this ever-expanding field of the biochemistry of the phosphatidylinositols and also for others laboring to understand what is happening in this area.

THE BIOCHEMICAL WONDERLAND.   The explosive growth in publications, in the last few years, on the phosphoinositides is directly attributable to seminal observations made over 40 years ago. Hokin and Hokin (1953) reported that membrane phospholipids turned over rapidly in slices of pigeon pancreas stimulated by carbamylcholine or acetylcholine. At that point in time, identi-fication of the involvement of any specific phospholipid was not provided. Nonetheless, it was an important observation. Only later was it shown that the inositol-containing phospholipids were intimately involved in the stimu-latory response. Early publications suggested that Ptd Ins was the major phospholipid in such reactions. Later, Michel et al. (1981) identified phosphatidylinositol-4,5-bisphosphate [Ptd Ins (4,5)-P_2] as the main phos-pholipid subject to turnover in stimulated cells. Furthermore, hydrolysis of the latter phospholipid was accompanied by Ca^{2+} mobilization. Important as these observations were at the time, another significant breakthrough occurred on identification of inositol-1,4,5-triphosphate as a Ca^2 mobilizing messenger (Streb et al., 1983). These reports provided important information on the role of a Ptd Ins $(4,5)P_2$-specific phospholipase C in stimulus-coupled responses.

During the surge in interest in the phosphatidylinositol derivatives, Takai et al. (1979) noted that an unsaturated diacylglycerol diminished the Ca^{2+} and phospholipid concentration required for complete activation of a Ca^{2+}-activated, phospholipid-dependent protein kinase (now known as protein kinase C). In this latter system, phosphatidylserine was most active (as the required phospholipid), with phosphatidylethanolamine and phosphatidyl inositol much less effective and phosphatidylcholine without any activity. Probably the enzyme was activated by an amphipathic molecule bearing a net negative charge, and any available cellular phospholipid mixture with the requisite surface change served this purpose.

On the basis of the above (and many other) findings, the current dogma states that the following series of reactions occur during cell stimulation:

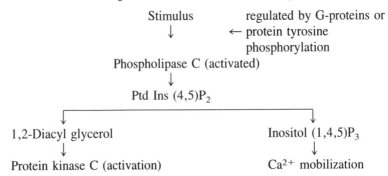

It should be emphasized that the diacylglycerol can be derived by phospholipase C action on other cellular phospholipids, but obviously the Ins $(1,4,5)P_3$ could derive only from Ptd Ins $(4,5)P_2$.

If further background material in this general area is desired, reviews by Nishizuka (1992) and by Michell, 1992 are highly recommend reading.

Isolation and Purification

Perhaps the best approach to extraction of the phosphatidylinositols and their various phosphorylated derivatives from a cellular preparation is through use of the Bligh–Dyer technique or some modification of it. However, it is of paramount importance that the chloroform–methanol–water (1:2:0.8, v/v) mixture, as an example, contain an acid, usually 1 N HCl as the water component. Otherwise, there will be a decreased recovery of the inositol phospholipids in the final chloroform extract. This is directly attributable to the fact that these inositol-containing phospholipids, as already mentioned above, are found naturally as the Ca^{2+}, Mg^{2+}, K^+, and/or Na^+ salts. If the solvent is not acidic, these salts essentially will remain in a water-rich fraction, whereas under acidic conditions, these compounds, in the free acid state, will be recovered in a chloroform-rich extract. The latter should be washed with a methanol–1 N HCl (1:0.8 v/v) mixture.

Subsequent to recovery of the total lipids of a cellular preparation as a chloroform-soluble fraction, the total phosphorus content can be determined (see Chapter 3); and then, depending on the amount of lipid phosphorus (or whether the preparation is radiolabeled or not, see below), analytical and/or preparative thin-layer chromatography can be undertaken. In either case, if the experimental protocol is centered on a signal-transduction process, then there may be insufficient material for a phosphorus analysis. In the latter instance, the cellular preparation is prelabeled with $^{32}P_i$ or [^3H]inositol and the labeled products are located by autoradiography. A preferred type of adsorbent (for thin-layer chromatography) is Merck silica gel 60 (oxalate impregnated). An effective solvent for separation of the phosphatidylinositols and other lipids is chloroform–acetone–methanol–acetic acid–water (80:30:26:24:14, v/v). The approximate R_f values of cellular phospholipids under these conditions are presented as follows:

Compound	R_f
Phosphatidylethanolamine, phosphatidic acid	0.90
Phosphatidylcholine	0.70
Phosphatidylinositol (Ptd lns)	0.55
Phosphatidylinositol-4-P[Ptd Ins (4) P]	0.30
Phosphatidyl inositol-4,5-bisphosphate [Ptd Ins (4,5) P_2]	0.24
Phosphatidylinositol-3,4,5-trisphosphate [Ptd Ins (3,4,5) P_3]	0.20

The more apolar lipids, such as triacylglycerols, cholesterol, and free fatty acids, would be located at the solvent front. If carefully performed, the above separation can be achieved on 250-μm as well as on 500-μm plates. Although chromatographic separations on the silica gel 60 can be agonizingly slow, they are usually excellent and very reproducible. Detection can take two forms, the first of which is autoradiography if the cells were labeled with radioactive precursors (where the usual procedure is to challenge the labeled cells with a particular agonist and examine changes in the labeled products). The second form of detection depends on use of a larger amount of material (not necessarily labeled) and locating the lipids through a spray such as TNS. In either case the desired areas are removed by scraping, extracted, and assayed. The latter will depend on the amount of material available for study.

COMMENTS ON SEPARATION PROCEDURES.   The tentative identification of the inositol phospholipids, especially those associated with the signal transduction pathway, is made possible through comparison of their behavior on HPLC columns or on thin-layer chromatograms with standard samples. There are synthetic or highly purified phosphatidylinositols and inositol phosphates available from commercial sources. While the R_f value of an unknown compound may agree with that of a "standard" compound, and while this may be a comforting factor, it still does not necessarily prove a chemical structure. In view of the small amounts of material usually available in a biological experiment, it is understandable that this approach is appealing and does have merit. However, it is very desirable to use additional modes of structure proof, and some of those outlined in the following section can be of value in many cases.

NOTES ON STEREOCHEMICAL CONFIGURATION. No attempt will be made here to undertake a comprehensive review of the stereochemistry of the inositol-containing phosphoglycerides. Nonetheless, it is well to point out certain facets of the nomenclature of these compounds that deserve some attention.

Throughout the history of study on these compounds, there have been many errors in nomenclature (particularly as it relates to their stereochemical notation). If a more serious consideration of this topic is desired, then one is referred to the very lucid, well-presented review of the subject by Parthasarathy and Eisenberg (1986).

However, it is still important to make a few comments here and lay a base for consideration of the conformation of the possible stereoisomeric forms of these compounds. Basically the primary form of the cyclohexitols found in mammalian cells is *myo*-inositol. This molecule contains a single axial hydroxyl at C-2, with all the other hydroxyls being equatorial. Due to the configuration of the C-2 and C-5 groups, which lie in the plane of symmetry, the molecule is considered achiral. It is occasionally referred to as the *meso* form. Configurational prefixes such as D- or L- cannot be applied to this molecule. Substitution of a phosphate, however, at certain but not all hydroxyls, can render it chiral. Finally the terms D- and L- are applied to illustrate the chirality of the molecule. The *myo*-inositol-1-phosphate derived from naturally occurring phosphatidylinositol (Ptd Ins) is the D-enantiomer. No L-isomer has been discovered in the inositol phospholipids of mammalian cells. However, this stereoisomeric form (L-) is produced in a biosynthetic pathway by isomerization of glucose-6-phosphate to L-inositol-1-phosphate, which is hydrolyzed to *myo*-inositol. In this presentation, the Haworth formulations are employed and serve to illustrate in a more basic manner the conformational characteristics of the molecule. However, the puckered (less strained) conformation for *myo*-inositol is widely used. As emphasized earlier, the reader again is urged to consult the classic review article by Parthasarathy and Eisenberg (1986).

Structure Proof and Analytical Detection

Positive identification of the various phosphatidylinositols can be frustrating due to the small amounts actually present in many cells. This latter point is amplified by the fact that certain of the derivatives are present in stimulated cells only and are at very low levels. Thus, as mentioned above, most of the analytical techniques employed for their assay depend on the use of radioactivity for location of the derivatives, comparison with "standard" inositol-containing compounds, and the use of a limited number of chemical structure proof procedures.

Basic proof of structure on the various naturally occurring phosphatidylinositols follow essentially the same protocol as outlined for the ethanolamine- and choline-containing phosphoglycerides. Primary attention here will focus, first, on an approach to proof of structure of phosphatidyl-*myo*-inositol, which is the major type found in mammalian cells, and can be isolated in milligram to gram quantities by preparative thin-layer chromatography. It will serve as a model compound, and the analytical techniques used to characterize it can be applied to structure proof studies on the other two main inositol containing phosphoglycerides, Ptd Ins (4)P and Ptd Ins $(4,5)P_2$. The latter can be obtained in milligram quantities from brain tissue. However, the recently discovered 3-phosphorylated [Ptd In-(3)P] and trisphosphorylated forms (Ptd In P_3) of phosphatidylinositol are present in very small amounts and hence present an impediment to an unequivocal structure proof.

SOME GENERAL PROCEDURES. Subsequent to isolation of this compound from thin-layer chromatographic plates (remember to use acidified solvent), the eluate can be subjected to the following assay procedures: total organic phosphate and fatty acid ester determinations. A phosphorus value in the range of 3.5–4.0% (based on a dry weight basis) and a fatty acid ester/P molar ratio of 2.0 should be obtained.

Evaluation of the apparent purity of the preparation can be gained from examination of its behavior on analytical thin-layer chromatography in two different solvent systems: *acidic*, chloroform–acetone–methanol–acetic acid–water (4.5:2:1:1.3:0.5, v/v) and *basic*, chloroform–methanol–28% ammonium hydroxide (65:35:6, v/v). Then, the use of the phosphorus, TNS, and ninhydrin sprays coupled with sulfuric acid plus charring will provide an insight into the uniqueness of the preparation. Phosphatidylinositol can be obtained as a white precipitate by treating a chloroform solution of it with 10 volumes of acetone. Centrifugation and removal of the supernatant will allow recovery of a white paste or solid (depending on the fatty acid composition). This residue is dried under a stream of nitrogen and can be stored in a tightly sealed container at $-25°C$ for several months.

STEREOCHEMICAL BEHAVIOR. In a solvent system of chloroform–methanol (10:1, v/v), native Ptd Ins will show an $[a]_D^{25} + 6.0$. Synthetic phosphatidylinositols, containing specific fatty acids and of a known *sn*-3 configuration, show a similar optical activity. The phosphatidylinositols with an *sn*-1 configuration will exhibit an $[\alpha]_D^{25}$ of $-5.5°$ to $-6.0°$. These results would be supportive of an *sn*-3 configuration for the native phosphatidylinositol.

Phospholipase A₂ Action.   This enzyme, obtained from *Naja Naja* or *Crotalus adamanteus* snake venom, will attack naturally occurring phosphatidylinositol. Using the ether-rich solvent system described in Chapter 4, phospholipase A_2 under proper conditions will cleave phosphatidylinositol completely with liberation of 1 mol of fatty acid (usually unsaturated) and 1 mol of *lyso*-phosphatidylinositol from 1 mol of starting material. These products can be isolated by thin-layer chromatography using an acidic solvent system. These results support an *sn*-3 configuration for the naturally occurring phosphatidylinositol.

Further Structural Analysis: Phospholipase C Attack.   Interestingly, not all sources of phospholipase C will attack phosphatidylinositol or its phosphorylated derivatives. However, a phospholipase C, which will attack all of these compounds, has been isolated from ram seminal vesicles by Majerus et al. (1986). This enzyme source will cleave Ptd Ins with formation of diglycerides and inositol phosphates. However, in addition to the expected inositol-1-phosphate, these investigators found a significant amount of an inositol-1,2-cyclic phosphate. In an earlier study, Dawson et al. (1971) reported the presence of an enzyme in pig thyroid tissue that would cleave phosphatidylinositol to diglyceride, inositol-1-phosphate, and inositol-1,2-cyclic phosphate. Although it is not clear as to the significance of the latter compound as applied to any cellular reaction, the results clearly show that a diglyceride is attached via a phosphate ester bond to the C-1 of the inositol molecule. Thus, the diglyceride was shown to have fatty acid esterified to the *sn*-1 and *sn*-2 positions of the glycerol backbone.

PREPARATION OF GLYCEROPHOSPHOINOSITOLS WITH CONCOMITANT FORMATION OF LONG-CHAIN FATTY ACID DERIVATIVES.   A primary approach used for structure proof and identification studies on the inositol-containing phosphoglycerides has been to effect the cleavage of the fatty acid ester groups and at the same time lead to the formation of glycerophosphoinositol. Base-catalyzed methanolysis of the parent molecule has been the procedure of choice.

The products of the base treatment reaction are recovered by phasing into chloroform and into water as described earlier in this and the previous chapter. The chloroform-soluble fraction will contain the methyl esters of the long-chain fatty acids, and the glycerophosphoinositols will be recovered in the water-rich fraction. It is important that the latter fraction be acidified and that the inorganic cations be removed by passage through an ion exchange column (e.g., Dowex-50).

The chemical characteristics and also the characterization of these products are discussed as follows:

Methyl Esters.   These esters can be isolated and purified by thin-layer chromatography on silica gel G with a solvent system of petroleum ether (b.p. 30–60°)–diethyl ether (80:20, v/v). Usually these esters exhibit an R_f value in the range of 0.80–0.85. Their characterization can be achieved in a manner similar to that described in Chapter 4. Thus, a combination of gas–liquid chromatography and mass spectrometry will afford a detailed insight into the structure and composition of these long-chain fatty acid esters.

The fatty acid composition of the phosphatidylinositols have some interesting features, particularly a high percentage of stearic acid residues (as compared to the other phosphoglycerides considered to this point). For example, in the phosphatidylinositol fraction obtained from human platelets, the 18:0 (stearic acid) species account for nearly 44% of the total fatty acid fraction, with the 16:0 chain length being only 2%. Even in the phosphatidylinositol isolated from sheep erythrocytes, the predominant saturated species is the 18:0 type (29 mol %), with the 16:0 type constituting approximately 10 mol %. On the other hand, the unsaturated fatty acid composition presents a more diverse profile. For example, in the phosphatidylinositol isolated from human platelets, arachidonic acid (20:4) forms 42 mol % of the total fatty acids. In the phosphatidylinositol obtained from sheep erythrocytes, there is no detectable arachidonic acid, with 18:1 (oleic acid) and 18:2 (linoleic acid) species accounting for nearly 40 mol % of the total fatty acids.

Perhaps the major take-home message here is that the fatty acid composition of the inositol phospholipids is quite different from that found in phosphoglycerides such as phosphatidylcholine or phosphatidylethanolamine. The uniqueness rests on the narrow spectrum of fatty acid chain lengths present.

Glycerophosphoinositol (GPI).   It is possible to isolate a highly purified phosphatidylinositol and from this preparation obtain a pure glycerophosphoinositol. It is important to remove cations from this preparation, and this can be achieved by use of ion exchange resins. If further purification is deemed necessary, this can be achieved by using high-pressure liquid chromatography with a Partisil SAX 10 anion exchange column as described by Irvine et al. (1985).

The GPI can be dissolved in water and analyzed for total phosphorus and for inositol. It should give an inositol/P molar ratio of 1.0. If a sufficient amount of the GPI is available, in the low milligram range, the sample can be neutralized with cyclohexylamine and then lyophilized. The residue is dissolved in a small volume of water, and approximately 10 volumes of acetone are added. A precipitate forms. This precipitate is dissolved in a small amount of water, and absolute ethanol is carefully added until a turbidity develops. The mixture is placed at −25°C for several hours, and the crystals are collected. These are washed with cold absolute ethanol and dried. This product is considered as the cyclohexylamine salt of glycerophosphoinositol.

Analysis of the above product for carbon, hydrogen, nitrogen, inositol, and phosphorus gave values corresponding to the cyclohexylamine derivative. It had a melting point of 124–126°C.

Periodic Acid Oxidation.   Treatment of the above derivative with periodic acid (in slight excess of theory for one *vic*-glycol) at room temperature showed a quantitative conversion of the sample to glycoaldehydephosphoinositol and formaldehyde (Brockerhoff and Hanahan, 1959). The results showed that the attack was only on the glycerol portion of the molecule because the attack of inositol, which must be done at an elevated temperature, would not yield formaldehyde. Thus, it was concluded that the glycerol was attached to the inositol molecule via a phosphate ester group.

While these results support a phosphate ester bridge between the glycerol and inositol, it does not prove where the bond is located on the inositol molecule. Inasmuch as the inositol phosphate isolated by alkaline cleavage from the parent phosphatidylinositol or the glycerophosphoinositol was optically active, the phosphate must be attached at the 1 or 4 position. Substitution at the 2 position of inositol would not yield an optically active product. Subsequently as will be discussed, observations from several laboratories have shown that the phosphate bond is between the *sn*-3 position of the glycerol and the C-1 hydroxyl on *myo*-inositol.

Some further comments on the periodic acid oxidation of GPI, as described earlier, have merit. The reaction is summarized in Figure 5-2. This reaction can be followed by the changes in optical activity, $[\alpha]_D^{25}$, values of the reactant and product. For example, the starting GPI shows a value of $-18.7°$, whereas the product glycoaldehydeinositol exhibits one of $+13.2°$. These results further support attachment of the phosphate at the 1 or 4 position. Interestingly, native phosphatidylinositol (PI) has a specific rotation value of $+5.5°$. Inasmuch as a synthetic dipalmitoyl (*sn*-3) phosphatidylinositol had a value of $+6.0°$, these obervations would support an *sn*-3 configuration for the native phosphatidylinositol.

Finally, periodic acid consumption can be followed as described in Chapter 4.

Inositol Determination.   Liberation of inositol from glycerophosphoinositol can be accomplished by acid hydrolysis, subjecting the sample to 2 N HCl at 125°C for 24–48 hr (preferably in a sealed tube). At the end of the

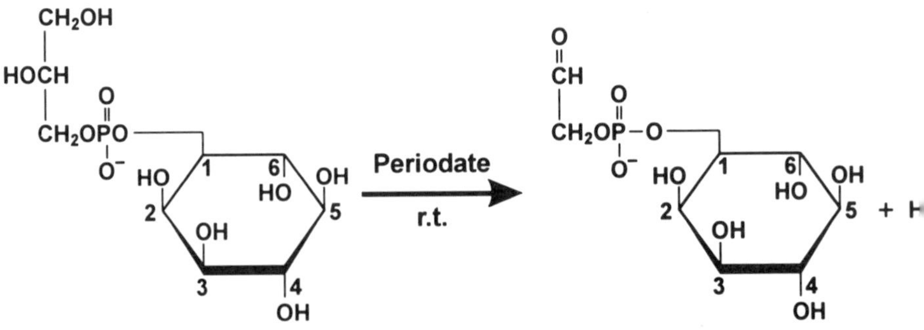

FIGURE 5-2. Periodic acid oxidation of glycerophosphoinositol (GPI).

reaction period, the tube is cooled to room temperature, and carefully opened, and the contents are extracted by the Bligh-Dyer technique. The water-soluble components are evaporated to dryness under nitrogen, and the residue is dissolved in water to volume. The analysis for inositol is conducted as described by Dawson and Freinkel (1961). This assay is a microbiological one, which uses *Kloeckera brevis*, an organism that requires *myo*-inositol for growth. The effective concentration range of this method is the range of 0.1–0.5 μmol inositol.

Inositol can be analyzed also by conversion to a silyl derivative. This entails reacting a free inositol sample with an *N,O*-bis(trimethylsilyl)trifluoro-acetamide–trimethylchlorosilane–pyridine (10:1:10, v/v) mixture. *Scyllo*-inositol is included as an internal standard. The derivatized sample can be analyzed by gas-liquid chromatography as described by Roberts (1987).

Inositol Phosphate Release and Identification. An effective route to re-lease of inositol phosphate from glycerophosphoinositol (GPI) is by alkaline treatment. The GPI is dissolved in 2 N KOH and heated at reflux for 30–45 min. The reaction mixture is cooled to room temperature and then acidified with formic acid. Subsequently this sample can be neutralized with ammonia and applied to an anion exchange column. A gradient elution with lithium chloride solution will elute the inositol monophosphate in the 0.2 M eluate, with any contaminating inositol diphosphate and inositol trisphosphate being eluted with 0.4 M lithium chloride (Grado and Ballou, 1961). A more sophis-ticated procedure using high-pressure liquid chromatography was developed by Dean and Moyer (1987) and can be used to separate several isomeric forms of the inositol phosphates. These two procedures can be used on a preparative basis as well as for identification of labeled products in a typical signal-transduction-type experiment.

Periodic Acid/Dimethylhydrazine Cleavage. Another approach favored by many investigators for recovery of inositol phosphate from GPI (as well as from GPIP, GPIP$_2$, etc.) is through use of periodic acid oxidation of the *vic*-glycol group (on the glycerol backbone) followed by dimethylhydrazine treat-ment to release the inositol phosphate. The technique was developed by Brown and Stewart (1966) and is summarized below:

$$\text{GPI} \xrightarrow[\substack{25°\text{C, 1–2 hr.}}]{\text{periodic acid}} \text{glycolaldehydeinositolphosphate}$$

$$\xrightarrow[\substack{25°\text{C}}]{\substack{4 \text{ hr,} \quad\quad \text{dimethylhydrazine}}}$$

Inositol monophosphate

This is a very smooth reaction, which can be used on milligram quantities of material or labeled products obtained in cellular experiments. The inositol phosphate can be isolated as the cyclohexylamine derivative as described next.

Cyclohexylamine Derivative.   Highly purified inositol monophosphate can be obtained by the above chromatographic procedures and then studied further. One effective approach is to mix the inositol phosphate with cyclohexylamine in a small volume (if milligram quantities are available, then the usual volume would be 1.5–2.0 ml). Acetone is added slowly to this clear solution until a turbidity is noted. Crystals will follow shortly (cooling to a low temperature can be helpful) and are collected by filtration or centrifugation. The crystals are washed with cold acetone and are then vacuum-dried and analyzed. Based on a molecular formula of $C_{18}H_{39}O_9N_2P$, an N value of 6.0% and a P value of 6.6% was obtained. The N/P molar ratio was 1.91 (theory, N, 6.1; P, 6.8; N/P, molar ratio 2.0). According to Ballou and Pizer (1960), this cyclohexylamine derivative, in water, will have an $[\alpha]_{589}^{25}$ of $-3.2°$ (at pH 9.1) and an $[\alpha]_{589}^{25}$ of $+9.3°$ (at pH 2.0). On the basis of comparison with synthetic compounds of known stereochemical configuration, this compound would be the D-myo-inositol-1-phosphate isomer.

A Summation.   It is possible at this point in time to isolate, separate, and identify glycerophosphoinositol and inositol monophosphate by sophisticated HPLC on anion exchange resins. This approach, using well-defined synthetic compounds (available from commercial suppliers) for standards, has made a significant impact on identification of inositol phospholipid metabolites in stimulated cell preparations. Certainly these latter techniques, together with the methodologies described earlier, have made life much easier for scientists in this field.

Characterization and Structure Proof of
Phosphatidylinositol-4-phosphate (PIP),
Phosphatidylinositol-1,4-bisphosphate (PIP₂), and Other
Phosphorylated Forms.

In the preceding section, attention was centered on phosphatidylinositol as a model compound and on its characterization and structure proof. In addition to PI, PIP, and PIP_2, additional forms have been discovered, and these indeed complicate the analytical scene. This is especially true if one is interested in investigating the role of these compounds in signal transduction processes. Unfortunately they are found in very small amounts, and radiolabeled precursors must be used in cellular preparation in which agonist responses are being studied. Nonetheless, the main thrust to their identification can be based on the following guidelines:

ISOLATION OF PARENT PHOSPHOLIPIDS.   This can be accomplished by using thin-layer chromatography with silica gel 60 as the adsorbent. The plate will have to be examined for its radioactivity profile. Inasmuch as many of these new phosphorylated forms of Ptd Ins migrate close together, the probability of overlapping species is significant.

DEACYLATION. The usual procedure is to deacylate these phospholipids in an alkaline medium and isolate the resulting glycerophosphoinositol phosphates by anionic exchange high-pressure liquid chromatography (HPLC). At the same time, the fatty acids released in the alkaline deacylation procedure can be investigated.

INOSITOL PHOSPHATES RECOVERY. A current trend in this field is to subject the particular glycerophosphoinositol phosphate to periodic acid oxidation coupled with dimethylhydrazine treatment. The liberated inositol phosphates then are subjected to anion exchange HPLC. Identification of the various inositol phosphates (by chemical as well as chromatographic methods) can be effectively accomplished by comparison to labeled as well as unlabeled synthetic derivatives available in high purity from commercial sources.

If one is interested in the specific facets of the above techniques and procedures, two recommended publications for reading would be that of Trayner-Kaplan et al. (1988) and that of Stephens et al. (1991). Further general information on the chemistry and biochemistry of inositol phosphates can be found in a (symposium) publication edited by Reitz (1991).

Phosphatidylserine: An Orphan Phospholipid

A discussion of the chemical as well as certain of the biochemical chracteristics of phosphatidylserine, whose structure is given in Figure 5-3, will serve as the final section devoted to the major non-choline-containing (acidic) phosphoglycerides. Even though phosphatidylserine can account for 4–14% of the membrane lipids of a mammalian cell, it can be best described as an "orphan phospholipid" due to its apparent lack of a role in cellular metabolism. Some attention will be paid later to the purported role of phosphatidylserine in cellular reactions; however, its critical importance must be viewed with some skepticism at present. Certainly it has not been implicated as a important component of any agonist-induced activation of a cell. Its importance in protein kinase C activity and in the blood coagulation process may be directly attributable to a physicochemical type of interaction. However, this topic will be explored in an abbreviated fashion later.

Notwithstanding its limited status in the biochemical behavior of a cell, phosphatidylserine is nevertheless an interesting compound. It is the only amino acid-containing phosphoglyceride found in mammalian cells. Furthermore, it exists only in the diacyl form, with no evidence to date for an alkylacyl or alkenylacyl form. Thus, it can be described as a 1,2-diacyl-*sn*-glycero-3-phospho-L-serine.

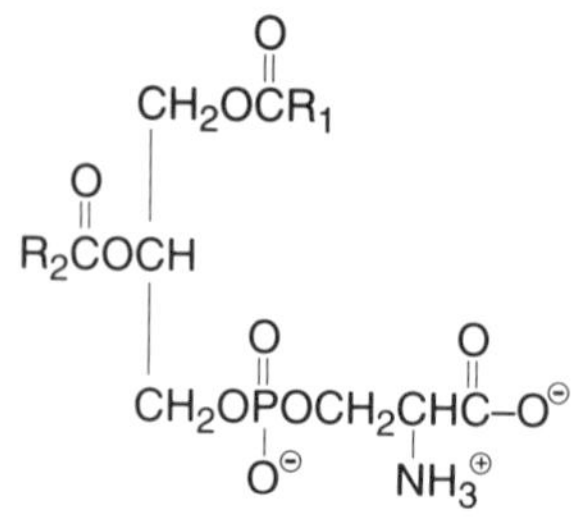

$$\begin{array}{c}
O \\
\| \\
CH_2OCR_1 \\
O \quad | \\
\| \quad | \\
R_2COCH \\
| \quad O \qquad O \\
| \quad \| \qquad \| \\
CH_2OPOCH_2CHC{-}O^{\ominus} \\
| \qquad | \\
O^{\ominus} \quad NH_3^{\oplus}
\end{array}$$

Phosphatidylserine
(sn-1,2-Diacyl-sn-glycero-3-phosphoserine)

FIGURE 5-3. Chemical structure of naturally occurring phosphatidylserine.

The following description of the isolation, purification and structure proof of a naturally occurring phosphatidylserine will be treated in exactly the same manner as with the previous phosphoglycerides. The methodologies described here can be used with any cell type and can be adapted for nearly any size sample. The exception, of course, would be if interest centered on the possible role of phosphatidylserine in a specific cellular reaction. Usually this would involve a small cell sample, and hence radioactive tracers would be needed. However, the biochemical pattern can be followed easily using exactly the same techniques.

Isolation

Two extraction procedures can be used to recover phospholipids from a biological sample, such as platelets, and both have their merits. A brief description of these procedures follows:

CHLOROFORM–METHANOL EXTRACTION.   Perhaps the most popular extraction technique in use today involves, in one form or the other, chloroform and methanol. The recovery of individual phospholipids such as phosphatidylserine is essentially quantitative. The total lipid sample is handled in exactly the same way as described for all the other cell lipids. Nonlipid contaminants can be removed by washing the final chloroform extract with a methanol–water (10:9, v/v) mixture. Phosphatidyserine is quite stable on storage, with little evidence for any lysophosphatidylserine formation or any oxidative changes in the fatty acid profile.

HEXANE–ISOPROPANOL EXTRACTION.   Hara and Radin (1978) introduced the use of a hexane–isopropanol mixture (3:2, v/v) as a solvent for extraction of lipids from a tissue sample. In their experiments on brain, 18 ml of solvent was used for each gram of sample. If nonlipid impurities are to be removed, washing of the hexane–isopropanol extract with aqueous sodium sulfate works well. At this stage, two phases are formed and the hexane-rich fraction contains all the lipids, except gangliosides, in good yield. The final total lipid extract can be analyzed for total phosphorus content and then subjected to further purification as described next.

The choice of the type of extraction technique, whether with chloroform–methanol or hexane–isopropanol, makes little difference in the total recovery of lipid. Removal of nonlipid contaminants can certainly be achieved with either of the aforementioned wash techniques. Thus the decision on the choice of the solvent for extraction rests solely with the investigator.

Purification and Recovery

Essentially three approaches can be used in the isolation of phosphatidylserine from a total lipid sample. These include thin-layer silica gel chromatography, aluminum oxide chromatography, and high-performance (high-pressure) liquid chromatography. The merits of these techniques are discussed as follows.

THIN-LAYER CHROMATOGRAPHY.   The total lipids can be applied in the usual way to a 20-cm × 20-cm silica gel H plate and run in the first dimension in a solvent system of chloroform–methanol–ammonium hydroxide (28%; 65:35:7, v/v). Then, it is run in the second dimension in a solvent system of acetone–chloroform–methanol–glacial acetic acid–water (48:36:12:12:5, v/v). The location of the phospholipid classes can be ascertained with an array of sprays, including ninhydrin, TNS, phosphate, and sulfuric acid char. In the usual thin-layer chromatographic run, a separate plate is spotted with a standard, well-defined phospholipid preparation and run in the same solvent systems in the same chromatography tank. This procedure is most useful for analytical purposes because preparative two-dimensional chromatography is not as successful or easily accomplished at higher sample loads. The main contaminant in most phosphatidylserine samples is phosphatidylinositol; this can be easily separated by the procedure just described. Finally this methodology is particularly advantageous in experiments utilizing radiolabeled compounds. Hence individual spots can be extracted and assayed.

ALUMINUM OXIDE CHROMATOGRAPHY.   This technique is particularly useful in rapidly separating neutral lipids and choline-containing phosphoglycerides from the non-choline-containing phosphoglycerides. Subsequently the ethanolamine- and serine-containing types can be separated from the inositol-containing phosphoglycerides. A brief description of this experimental protocol follows.

Aluminum oxide [BioRad Alumina (oxide), AG-7] is suspended in chloroform–methanol (1:1, v/v) and then transferred to a glass chromatography tube fitted with a Teflon stopcock. Solvent is allowed to flow through until the adsorbent has settled to a constant height. The recommended height-to-diameter ratio is 10, and the loading ratio is 0.5–1.0 mg of lipid P per gram of aluminum oxide. The sample is dissolved in a small volume of chloroform–methanol (1:1, v/v) and applied carefully to the top of the column bed. After the solvent has nearly disappeared into the column, small samples of the initial solvent are added sequentially to wash the sample completely onto the column.

Additional chloroform–methanol (1:1, v/v) is allowed to pass through the column (with slight nitrogen pressure to accelerate the passage). The eluate is monitored continually for lipid P and for sulfuric acid charrable material. Examination of this total eluate will reveal that it contains all the neutral lipids plus the choline-containing phospholipids. When the eluate shows the absence of detectable lipid P and material positively to charring, the solvent is changed to ethanol–chloroform–water (5:2:1, v/v), and the eluate is monitored in the same way as above. This eluate will contain phosphatidylethanolamine and phosphatidylserine. In this solvent system the elution tends to slow down, and it is necessary to apply about 5 psi nitrogen pressure to aid in the solvent flow. If the column were further treated with ethanol–chloroform–water (5:2:2, v/v), phosphatidylinositol will be eluted.

The ethanol–chloroform–water (5:2:1, v/v) eluate is evaporated to dryness *in vacuo* and the residue is dissolved in chloroform–methanol (2:1, v/v). This fraction can be analyzed for total P, and its contents can be evaluated by thin-layer chromatography on silica gel H (250 μm) plates in a solvent system of chloroform–methanol–ammonium hydroxide (28%; 65:35:7, v/v). Two ninhydrin-positive spots will be found: one is at R_f 0.50, which is phosphatidylethanolamine; the other is at R_f 0.20, which is phosphatidylserine. These findings then make it possible to isolate phosphatidylserine by preparative thin-layer chromatography.

HIGH-PERFORMANCE (HIGH-PRESSURE) LIQUID CHROMATOGRAPHY. This technique was used successfully by Dugan et al. (1986), to separate all the major classes of lipids obtained from a bovine brain plasma membrane. The column material was a DuPont Zorbax SIL, a silicaceous material, and the eluting solvent system was hexane-2-propanol (3:2, v/v). An amount of phospholipid equivalent to 180 nmol lipid P could be separated on a 25-cm × 4.6-mm column in approximately 40 min. Elution of individual peaks was monitored by a detector at 205 nm. Essentially phosphatidylserine could be separated quite well from phosphatidylinositol and lysophosphatidylethanolamine. A comparison of the results (using lipid P assay) with the high-performance (high-pressure) liquid chromatography run and two-dimensional thin-layer chromatography showed excellent correlation. Though not explored as a preparative procedure, it seems likely that high-performance liquid chromatography could be adopted to larger amounts. The speed of this latter methodology certainly is an attractive feature.

Proof of Structure

Assuming at this point that a highly purified sample of phosphatidylserine has been obtained from a specific biological source, structural chracterization of this preparation can be undertaken with ease and confidence. Prior to delving into the latter topic, a few general comments are appropriate at this time.

SOME GENERAL COMMENTS.   The highly purified phosphatidylserine sample should migrate as a single spot (or band) on thin-layer chromatography in a basic and an acidic solvent system. This behavior can be checked by one-dimensional as well as two-dimensional thin-layer chromatography. Assay of the phosphatidylserine sample for total P and total N should give an N/P molar ratio of 1.0. A further assay for fatty acid ester content (see Chapter 4) will show a fatty acid ester/P molar ratio of 2.0. In normally encountered phosphatidylserine samples from mammalian cells, there is no detectable amount of the alkenylacyl and/or alkylacyl analogs present. Armed with this information, a more rigorous definition of the structure of this phospho-glyceride can be undertaken.

QUANTITATIVE ASSAY.   The estimation of the serine-containing (as well as the ethanolamine-containing) phosphoglyceride levels in mammalian cell lip-id extracts can be accomplished quantitatively through use of a fluorescent probe procedure described by Chen et al. (1983). In this assay the amino-containing phosphoglycerides are treated with succinimidyl-2-naphthoxyacetate and the products then separated by high-performance liquid chromatogra-phy on a silica gel column. Using a gradient elution technique, the progress of the separation can be monitored by examination of the eluates at 228 nm (excitation) and 342 nm (emission). As little as 25 μg of total phospholipid can be derivatized successfully and analyzed. This method is simple, fast, and reproducible. An added value of this analytical procedure is that lysophos-phatidylserine (and also lysophosphatidylethanolamine) are well separated from their diacyl counterparts and can be quantitated also.

CHARGE CHARACTERISTICS OF POLAR HEAD GROUP.   It should be obvious from the above structural formula that phosphatidylserine possesses three ionizable groups, namely, a diester phosphoric acid, an α-amino group, and a carboxyl function. Hence, it should possess different pK values. In titrimetric studies on this compound dissolved in 2-ethoxyethanol, Garvin and Kar-novsky (1956), showed, as expected, that there were three available ionizable groups. The data supported the conclusion that at pH 7.0 the phosphate and the carboxyl functions were in the anionic form and that the α-amino group was protonated (i.e., positively charged). Interestingly, phosphatidylserine as normally isolated from mammalian cells contains 1 mol of a cation, usually K^+ or Na^+, bound presumably to the carboxyl function (although an interaction with the phosphate group could not be excluded). These cations can be removed by washing the phosphatidylserine (in chloroform) with 0.05 N HCl.

BASE-CATALYZED METHANOLYSIS.   Using exactly the same conditions as described earlier in this chapter and in Chapter 4, the cleavage of phosphatidylserine by 0.5 M NaOH in methanol proceeds smoothly to completion within 15–20 min at room temperature. As expected, the products are the methyl esters of long-chain fatty acids and glycerophosphoserine. The methyl esters are easily recovered into chloroform, and the glycerophosphoserine is found as a water-soluble component. Some characteristics of these two components will be considered next.

Methyl Esters.   This fraction can be examined first by its behavior on unidimensional thin-layer chromatography. In a solvent system of petroleum ether (30–60°)–diethyl ether–glacial acetic acid (90:10:1, v/v) and using silica gel G (250 μm) plates, there was only a single spot at R_f 0.65–0.67. This compared exactly with a standard synthetic methyl palmitate. As described in Chapter 4, the chemical nature of the methyl esters can be obtained by analysis on gas-liquid chromatography coupled with mass spectrometry.

The latter analytical procedure provides an interesting insight into a rather restricted range of long-chain fatty acids encountered in mammalian cell phosphatidylserine. For example, over 97 mol % of the long-chain fatty acids found in human platelet phosphatidylserine was represented by 18:0, 18:1, and 20:4 chain lengths. Of further interest, approximately 45 mol % of the chain lengths is the 18:0 species. This type of distribution appears not to be unusual because the fatty acids in the phosphatidylserine of human erythrocytes are composed of nearly 40 mol % 18:0 chain length. In the latter source, over 90 mol % of the fatty acids are represented by 18:0, 18:1, 20:4, and 22:6 chain lengths. For only a brief comparison, the fatty acids in the phosphatidylethanolamine fractions in other cells, the 18:0 chain length represents only 8 mol % of the total, whereas the 16:0 species comprised 25 mol %. These data illustrate dramatically there there is a significant specificity in handling (and incorporation) of fatty acids, by chain length, in mammalian cells.

Glycerophosphoserine.   The water-soluble fraction from the above procedure should contain only *sn*-glycero-3-phospho-L-serine. Analysis for total lipid P and total N should give an N/P molar ratio of 1.0. In addition to these assays, proof for the presence of serine is very important and can be proven by the analytical procedure(s) described later in this section. A serine (N)/P molar ratio of 1.0 would be expected.

Vic-Glycol Content.   This can be determined through use of the periodate cleavage technique. This methodology has been described earlier and with glycrophosphoserine, there will be 1 mol periodic acid consumed (per mole of

P) and concomitant liberation of 1 mol of formaldehyde. This result proves that there is a *vic*-glycol present, but it does not prove whether it has an *sn*-3 or an *sn*-1 configuration. Obviously, an *sn*-2 configuration is eliminated by the periodic acid cleavage experiments.

STEREOCHEMICAL CONFIGURATION.   It is possible to establish the configuration of the original PS sample in two ways, one being by its optical rotation value compared to synthetic samples and the other by the nature of its response to phospholipase A_2.

Optical Rotation.   A reasonable approach to the definition of the stereochemistry of the naturally occurring phosphatidylserine is to compare its optical activity with that of a known stereochemical structure. A synthetic *sn*-3 distearoylphosphatidylserine has an $[\alpha]_D^{25}$ of 16.2, whereas a naturally occurring PS sample (e.g., obtained from human platelets) has an $[\alpha]_D^{25}$ of 15.8. This observation would support an *sn*-3 configuration for the naturally-ocurring material.

Phospholipase A_2 Action.   Incubation of phosphatidylserine with phospholipase A_2 obtained from *Crotalus adamanteus* or *Naja Naja* snake venom will show that the serine-containing phosphoglyceride was smoothly and completely converted to a lysophosphatidylserine with liberation of 1 mol of fatty acid per mole of lipid P. The experimental procedure was the same as the one described before in this and in the previous chapter. The products of the reaction can be recovered by thin-layer chromatography on Whatman K6 plates in a solvent system of chloroform–acetone–methanol–acetic acid–water (4.5:2:1:1.3:0.5, v/v).

The lysoPS migrates to an R_f of 0.2, and the free fatty acid migrates to an R_f value of 0.70. These compounds can be isolated from the plates and studied further as desired.

These results support the assignment of an *sn*-3 configuration to the naturally occurring phosphatidylserine. A synthetic *sn*-3-phosphatidylserine sample is easily attacked, whereas an *sn*-1 phosphatidylserine is not.

ISOLATION AND IDENTIFICATION OF SERINE.   It is important to establish that serine is present in the originally isolated sample and then to prove that it is of the L or D stereochemical configuration. This substituent can be cleaved from a phosphatidylserine sample by an acid hydrolytic procedure. A sample as small as 100 μg total weight is suspended in 6 N HCl, and the mixture is heated at reflux for 60 min. After cooling to room temperature, the fatty acids are removed by filtration and the water-soluble fraction is evaporated to dryness *in vacuo*. The residue is dissolved in 2–3 ml distilled water (if any color is present, it can be removed by charcoal treatment). To the clear, colorless solution is added 30–35 mg of *p*-hydroxyazobenzene *p*-sulfonic acid, and the mixture is warmed to effect solubilization of the sulfonic acid. Storage of the clear solution at 4°C will result in excellent crystal formation within two days. The crystals are recovered by centrifugation at 4°C and washed twice with cold water. The original supernatant and the washings are combined and evaporated to 2–3 ml and allowed to stand at 4°C. Additional crystals form and are collected, and then they are combined with the first crop. The yield is normally in the range of 80–85%. This derivative exhibits an optical activity $[\alpha]_D^{25} + 14.3°$ (c, 10; in 1 N HCl). Carbon, hydrogen, and nitrogen analyses can be run and compared with a synthetic L-serine derivative. The optical activity for the D-serine complex would be exactly the opposite of the L-serine derivative.

The serine sulfonate complex can be treated with lead acetate, the insoluble lead salt can be removed by filtration, and the water-soluble fraction can be recovered. The latter is evaporated to dryness and dissolved in 1 ml of distilled water. To this clear solution is added 10 ml of ethanol, and the mixture is stored at 4°C. Crystals form and are collected by centrifugation and dried *in vacuo* at room temperature. They have the following characteristics: $[\alpha]_D^{25} + 14.7°$ (c, 10, in 1 N HCl); m.p. 220°C (with decomposition). These values compare very favorably with a synthetic L-serine. Synthetic D-serine has an $[\alpha]_D^{25} - 14.4°$.

An alternative route to isolation of serine is to eliminate the above sulfonic acid reation and simply crystallize the serine from the hydrolysate (filtered) by the addition of ethanol. The recoveries are not as good as with the above derivative procedure, and the latter also provides additional proof for the presence of serine in the phosphoglyceride preparation.

Finally, a very sensitive and easily handled fluorescence method for the qualitative identification of serine involves thin-layer chromatography of its dansyl derivative on "polyamide" sheets. The sample for assay is reacted with commerically available dansyl-Cl reagent and then subjected to thin-layer chromatography on polyester sheets bonded to ε-polycaprolactam. The separation is undertaken in a two-dimensional mode. In the first direction, water–formic acid (9:1, v/v) is used and the serine derivative migrates to an R_f 0.85–0.90. The solvent for the second direction is benzene–glacial acetic acid (9:1, v/v), and serine migrates to an R_f 0.25–0.30. A separate chromatogram should be run with pure serine alone. Then another run with a mixture of threonine, arginine, and glycine would be advisable. The DNP-amino acids are located by exposure of the plate to ultraviolet light. This technique can detect as little as 1 μg of serine.

QUANTITATIVE ASSESSMENT OF SERINE. A casual reading of the literature devoted to analysis of peptides and proteins will attest to the plethora of approaches to determination of their amino acid composition. A review of this field is well presented by Hunkapillar et al. (1984), which is recommended reading for anyone contemplating a study of the amino acid composition of a lipid.

Among the many methods described for quantitative determination of serine, the one reported by Adams (1974), provides a simple and rapid assay for serine. It involves gas chromatography of the *n*-propyl, diacetyl derivative of serine. The initial derivatization steps are accomplished easily within 25 min, and the subsequent gas chromatography can be done within 15 min. The detection limit is at the nanomole level. Recoveries are 90% or better, and the instrumentation is within the capabilities of most research laboratories. If more detailed information is desired on the derivatives(s), a combined gas chromatography–mass spectrometry unit would be of great value. An alternative route is to use high-performance liquid chromatography. It all depends on the direction of the research program and the funds available for the latter sophisticated procedure.

PHOSPHOLIPASE C AND PHOSPHOLIPASE D AS REAGENTS FOR STRUCTURE PROOF. Convincing support for the proposed structure of phosphatidylserine can be obtained through the use of phospholipases C and D. Each of these enzymes attacks on opposite sides of a phosphodiester bond in a phosphoglycerides. In each case, two fragments are formed and are of great value in confirming the basic structure of a phosphoglyceride. In the one instance, phospholipase C action will yield *O*-phosphoserine and a diacylglycerol; in the other instance, phospholipase D action will produce a phosphatidic acid and free serine. Some additional notes on these reactions follow.

Phospholipase C. Using a phospholipase C obtained from *Bacillus cereus*, the experimental conditions for attack on phosphatidylserine are exactly the same as those described in Chapter 4. The reaction proceeds smoothly and leads to a chloroform-soluble, phosphorus-free product (i.e., diacylglycerol) and a water soluble phosphorus-containing product [i.e., *O*-phosphoserine (*O*-PS]).

The chloroform-soluble product can be analyzed first by thin-layer chromatography (in comparison to a known diacylglycerol) and then by gas liquid–mass spectrometry as described in Chapter 4. The water-soluble fraction should contain all of the original phosphorus in an organic form, and all of the serine should be covalently bound to the phosphate. Analysis of this fraction for total phosphorus and for serine should give a serine/phosphorus molar ratio of 1.0. As further proof on the structure of *O*-PS, the water-soluble fraction can be converted into its *N*-isobutoxy carbonyl methyl ester derivative. The latter material can be analyzed by gas chromatography with flame ionization detection, and the *N*-isobutoxy derivative is well separated from *O*-phosphoethanolamine. The latter compound is derived from phosphatidyethanolamine and would form an *N*-isobutoxy carbonyl methyl ester also. Use of standard *O*-phosphoserine and *O*-phosphoethanolamine samples (easily available from commerical sources) for comparison are highly recommended. An added check on the structure of these products can be gained by gas chromatography–mass spectrometry. The details of this structure proof approach are described in a paper by Kataoka et al. (1989).

The results of the above experimental approach will provide excellent proof for an *sn*-1,2-diacylglycerol backbone attached by a phosphodiester linkage to O-phosphoserine. Thus the original sample is 1,2-diacycl-*sn*-glycero-3-phosphoserine. The proof for assignment of an L configuration to the serine moiety is provided earlier in the section entitled "Isolation and Identification of Serine."

Phospholipase D.   This enzyme will attack phosphatidylserine with the liberation of serine and formation of phosphatidic acid. The methodology is exactly the same as the one outlined in Chapter 4. The source of enzyme can be *Streptomyces chromofuscus* or cabbage, and products of its action are recovered in a chloroform-soluble and a water-soluble fraction. All of the lipid P should be in the chloroform-soluble fraction, and all of the serine should be in the water-soluble fraction. The phosphatidic acid can be identified by its thin-layer chromatographic behavior and its fast atom bombardment–mass spectrometric pattern. Serine can be identified by the procedures outlined earlier.

SYNTHESIS OF PHOSPHATIDYLSERINE.   Though there have been strictly organic chemical procedures reported for the synthesis of phosphatidylserine, they usually give low yields and are arduous at best. A much smoother, more facile, and faster approach is an enzymatic one where the trans-phosphatidylase activity of phospholipase D is utilized. Essentially this involves interaction of phospholipase D with a well-defined, high-purity phosphatidylcholine preparation in the presence of L-serine. In a study in which various sources of phospholipase D were employed, Juneja et al. (1989), described in careful detail the conditions leading to formation of phosphatidylserine in a 97% yield. Inasmuch as there are several different types of phosphatidylcholine (with varying, but well-defined fatty acid composition) available from commerical sources, this biosynthetic route is the one of choice.

MASS SPECTROMETRIC EVALUATION.   A limited amount of information is available on the mass spectrometric examination of phosphatidylserine. Recently, however, a study was reported by Kerwin et al. (1994) on the analysis of phosphatidylserine by electrospray ionization mass spectrometry. Interestingly, at a concentration of 2 ng/ml, spectra were obtained within a total acquisition time of 10 sec. Prominent species were [M + H$^+$] and [M + Na$^+$] adducts, a fragment of mass 87 for serine and one at 185 (208 if Na$^+$ present) for O-phosphoserine.

Mass spectrometry as applied to simple and complex lipids has been an invaluable aid in establishing the structural features of these molecules. There is a drawback, however, for the general research community, and that rests solely on the sophistication (and cost) of the instrumentation. In addition, one needs access to a talented mass spectrometrist capable for interpreting the acquired spectra. If these two barriers can be overcome, then researchers even casually involved in lipid chemical experimentation are advised strongly to take advantage of this impressive methodology.

INVOLVEMENT OF PHOSPHATIDYLSERINE IN BIOLOGICAL REACTIONS. As mentioned earlier, phosphatidylserine has not been shown to be actively involved in the signal transduction system or mammalian cells. However, this phosphoglyceride has been studied in great detail as a key component of protein kinase C activity and in certain facets of the blood coagulation process. A brief consideration of each of these topics is outlined as follows.

Cytosolic protein kinase C requires both Ca^{2+} and an acidic phospholipid for optimum *in vitro* (kinase) activity. Phosphatidylserine has been the phosphoglyceride of choice, and models have been constructed describing the importance of electrostatic interactions in the binding of phosphatidylserine to the kinase (or to a membrane). Also, it is suggested that the selective binding of phosphatidylserine in a cooperative manner is most important to the efficient formation of a fully activated complex. Much effort has been expended by many laboratories to prove the uniqueness of the phosphatidylserine effect. However, it would be of value to all investigators in this area to read a review by Nishizuka (1992), in which he explores the importance of other pathways (and lipids) to the activation of protein kinase C.

An important step in the blood coagulation pathway is the formation of the prothrombinase complex. The latter is a mixture of factor V, factor X_a, Ca^{2+}, and phospholipid. In this case, a phospholipid mixture with a net negative charge will allow the prothrombinase complex to form. This active enzyme is important in cleaving prothrombin to yield thrombin. The most active phospholipid mixture for *in vitro* studies has proven to be phosphatidylserine-phosphatidylcholine. Subsequently the hypothesis has developed that phosphatidylserine is key to the formation of prothrombinase.

There is no doubt that phosphatidylserine is active in the above reactions. However, in neither of these two systems has the importance of this phosphoglyceride in an *in vivo* system been defined. Certainly in the protein kinase C studies, there appeared to be no real attempt to explore the effect of mixtures of phospholipids (which might be encountered in the cell) on the kinase activity. In the blood coagulation pathway, it is well established that phosphatidylserine-phosphatidylcholine mixtures with the proper net negative charge are key in the formation of the active prothrombinase. Consequently, there is a general assumption by many investigators in this field that phosphatidylserine is the key component.

It may well be correct that phosphatidylserine is integral to the reactions described. However, it is only a hope that at some point in the future it will be proven whether this is true for the *in vivo* situation. It will be of equal importance to establish *in vivo* whether mixtures of phosphoglycerides have an importance also. These are complex systems and it is obvious that further investigation is required to define the role of phosphatidylserine in these biologically important reactions.

A FINAL TRANSITION.　The information presented in this and previous chapters should provide a comfortable basis for any experimental inquiry directed toward the chemistry and biochemistry of mammalian phospholipids in particular. It should be emphasized again that this book is intended expressly as an introduction to a fascinating field of study and not as a tome. It seems fitting to conclude this presentation with a discussion of phospholipids, which exhibit distinct biological properties or have a unique structural status as cellular phospholipids.

This area of research investigation has developed explosively in the past 15 years, and it is hoped that the final chapter will provide a sense of excitement in this field.

SIX

MINOR PHOSPHOLIPIDS

Platelet Activating Factor (PAF) and PAF Analogs,
Lysophosphatidic Acid and Phosphatidic Acid,
Phosphatidylglycerol Cardiolipin,
Sphingosine-1-P

An Exciting Vista

In the previous chapters, the emphasis was placed on establishing the chemical nature of phospholipids normally found in relatively high concentrations in mammalian cells. The methodology described for identification of these compounds can be applied, perhaps with some minor modifications to the structural characterization of phospholipids found in plants, fish, and bacteria. Interwoven into the fabric of these chapters was the subtle reference to their potential biological role in cellular behavior. As lipid biochemistry has developed over the past several years, there is no doubt that the field of signal transduction has had an enormous impact on phospholipid awareness. As might be expected, the potential for phospholipids or their metabolites to exhibit biological activity loomed large.

As will be evident in this chapter, phospholipids can certainly have biological activity and this theme shall be explored in some depth. The title "Minor Phospholipids" is not meant to be belittling, but is only intended to reflect the fact that these phospholipids are present in very low concentrations. In fact, some are considered to be formed only after a cell has been stimulated by an extracellular agonist. As shall be seen, these compounds have considerable biological activity and hence are really of major importance in the cellular metabolic scene.

Platelet Activating Factor (PAF) and PAF Analogs

Platelet Activating Factor

General Comments

The term *platelet activating factor* (PAF) was originally applied to a phospho-glyceride capable of activating platelets, leading to their aggregation. In addition, in certain species, there was also discharge of their dense granules (as indicated by serotonin release from the platelets). This nomenclature was unfortunate because it is now well established that many cells can produce this compound and that, likewise, PAF can stimulate many other cells. There are many other compounds, such as thrombin and arachidonic acid, which also can activate platelets. Notwithstanding this problem of nomenclature, there is a widespread (deeply entrenched) usage of this term to indicate a specific type of phospholipid with a particular biological activity. Given the status of this field at the current time, the use of the term PAF will be continued here.

Platelet activating factor can be defined as a phosphoglyceride with potent inflammatory properties as well as many other physiological and pathological attributes. The most prevalent chemical form of this factor can be represented by the structural formula shown in Figure 6-1, where n = 15:0, 17:0, 17:1 (in highest amounts). Its chemical name is 1-*O*-alkyl-2-acetyl-*sn*-glycero-3-phosphocholine.

In addition to the above form of PAF activity isolated from such cells as stimulated platelets, neutrophils, and liver, an *sn*-1 fatty acyl analog, with much lower biological activity, has been detected in certain cells, and its structure has been well proven to be 1-*O*-long-chain fatty acyl-2-acetyl-*sn*-glycero-3-phosphocholine. Interestingly, these *sn*-1 fatty acyl analogs of PAF have been shown to exist also as the proprionyl, acryoyl, butyryl, valeryl, caproyl, and heptanoyl-*sn*-2 derivatives. While it is well established that the most potent PAF activity is associated with the *sn*-1-alkyl and *sn*-2-acetyl derivative, the occurrence of a family of *sn*-1-long-chain fatty acyl–*sn*-2-

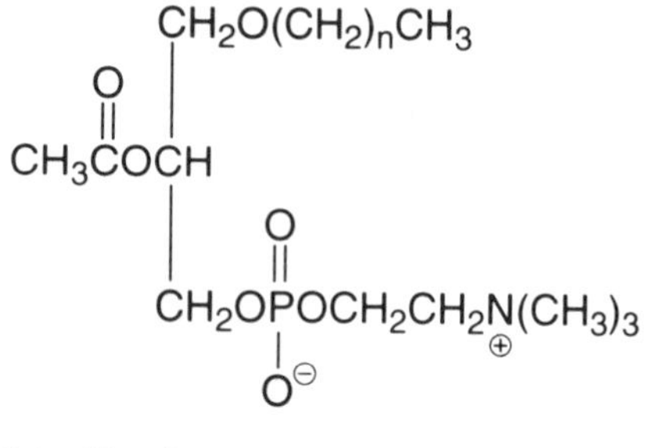

$$CH_2O(CH_2)_nCH_3$$

$$CH_3\overset{O}{\overset{\|}{C}}O\overset{|}{C}H$$

$$CH_2O\overset{O}{\overset{\|}{P}}OCH_2CH_2\overset{\oplus}{N}(CH_3)_3$$
$$\overset{|}{O}_{\ominus}$$

FIGURE 6-1. Chemical structure of platelet activating factor.

short-chain acyl forms demands close attention (Tokumura et al., 1989). The importance and role of the latter, naturally occurring forms awaits further detailed investigation.

A further exploration of the composition of the PAF-rich fraction from certain cells revealed, not unexpectedly, the presence of the vinyl ether (plasmalogen) analog of PAF. Nakayama and Saito (1989) reported the occurrence of 1-O-alk-1'-enyl-2-O-acetylglycerophosphocholine in perfused rat and guinea pig hearts. The major species of this vinyl ether derivative was recovered by reversed-phase high-performance liquid chromatography (HPLC). This technique allowed the separation of the alkenyl from the alkyl ether forms. Structural characterization was achieved by mass spectrometry together with selected ion monitoring (SIM). Interestingly, Tessner and Wykle (1987) showed that stimulated human neutrophils produced an ethanolamine-containing vinyl ether analog of PAF, namely, 1-alk-1'-enyl-2-O-acetyl glycerophosphoethanolamine. Under the same experimental conditions, a much smaller amount of the 1-O-alkyl ethanolamine analog was formed. The biological significance of these vinyl ether derivatives remains to be established.

An enormous number of publications has emerged over the past 16 years on various aspects of PAF chemistry, biochemistry, and physiology. Certainly the field has expanded tremendously from the days when its behavior on rabbit platelets was a major experimental outlet. Currently the biological activities of PAF (the alkyl form) can be separated into the following two classes:

INFLAMMATORY AND IMMUNE RESPONSES AND RESPIRATORY, CARDIO-VASCULAR, REPRODUCTIVE, AND NERVOUS SYSTEM PHYSIOLOGY. In all of these systems, the effects are mediated through a receptor(s) located on the plasma(s) membrane of responsive cells. Usually the biological effects noted are directly associated with stimulated cells or tissues. There is also support for the presence of PAF (and certain analogs) in normal tissues, with examples being the brain, uterus, lung, and the glandular stomach.

As might be expected in a dynamic area of research, a number of broad-spectrum reviews have appeared. Other than for brevity's sake, only the following are suggested: Hanahan (1986); Venable et al. (1993); Shimizu et al. (1992); Saito (1992).

In the succeeding sections, the major emphasis will be directed toward the characterization of naturally occurring phosphoglycerides with PAF activity and its analogs.

Isolation and Recovery

In the usual experimental protocols involving PAF, the objective often is to determine whether PAF is formed under stimulatory conditions or whether it has been metabolized by cells, cellular extracts, or specific enzymes. In most instances, the reactions can be followed only through the use of radiolabeled precursors and certain simple analytical procedures. The methodology to be described here can be applied, with care, to the above type of studies. More refined techniques, involving derivatization and mass spectrometry, will be discussed later. If the experiments involve milligram quantities of PAF or analogous compounds, then the characterization really is very straightforward and easy.

The isolation and purification of PAF from a cellular preparation (e.g., platelets, neutrophils, Kupffer cells) follows exactly the same procedures as described in Chapter 4 for the choline-containing phosphoglycerides. Essentially the solvents of choice are chloroform and methanol, with the ultimate aim of recovery of the PAF in a chloroform-rich fraction. Certainly the most effective and most facile route to separation of the lipid extract into distinct fractions is by thin-layer chromatography. The type of adsorbent, the solvent systems, and the general detection reagents are exactly the same as outlined in Chapter 4 (and also reiterated in Chapter 5). There is only one additional very important modification to the thin-layer chromatographic procedure. Usually the amount of PAF formed in these experiments is insufficient for detection by reagents such as the phosphorus or fluorescent sprays. It is mandatory, therefore, to run a control lane containing synthetic PAF (sufficient for detection by spray reagents) or a tritiated PAF and also lysophosphatidylcholine and sphingomyelin. This lane should be two to three lanes away from the experimental lipid sample so that any possible cross-contamination (from the standard PAF) can be avoided. Subsequent to running the chromatogram, the lane containing the experimental sample is removed by scraping 0.5-cm sections starting from the origin to the solvent front. Later, depending on the results, the number of sections to be scraped can be modified. Then these individual sections can be extracted with chloroform–methanol–water (1:2:0.8, v/v), the extract can be phased per usual, and the chloroform-soluble components can then be assayed as described below. As a check on the chromatographic separation of the lipids, an additional lane spotted with the experimental sample can be run at the same time but not scraped. Then after scraping of the other experimental lane, the plate can be sprayed with the TNS or phosphorus reagent and the R_f values compared with those of the lane with the standard samples.

In a typical thin-layer chromatographic separation of the lipids from a stimulated cell, the profile shown in Figure 6-2 can be obtained. The thin-layer chromatographic plate in this case is coated with silica gel G, and the solvent system is chloroform–methanol–water (65:35:7, v/v). Not all of the lipids are shown on this chromatogram because the main attention is centered on the position of PAF. The abbreviations used in the figure are as follows: PE, phosphatidylethanolamine; PC, phosphatidylcholine, S, sphingomyelin; PAF, platelet activating factor; Lyso PC, lysophosphatidylcholine; NL, glycerides, cholesterol, and free fatty acids; SF, solvent front; O, origin.

The PAF activity is located by a biological assay. On occasion, perhaps due to loading factors and type of adsorbent, the PAF activity may be found, in part, in the sphingomyelin section. However, this poses no problem because sphingomyelin has no demonstrable activity toward platelets. However, if the

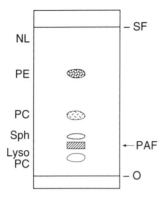

FIGURE 6-2. Migration behavior of platelet activating factor on a thin-layer chromatogram.

desire is to have a "clean" preparation of PAF, chromatography of the PAF-rich fraction on a silica gel plate in a solvent system of methanol–water (2:1, v/v) will allow a very adequate purification of the PAF (Satouchi et al., 1981). Under these conditions, sphingomyelin remains at the origin whereas PAF migrates to R_f 0.45.

An alternative, but more expensive, means of location of PAF on a chromatogram is to include [^{3}H]PAF in the lane with other standard phosphoglycerides. Then at the end of the chromatographic separation, the plate is air-dried and examined using a BioScan unit as described in Chapter 4.

Biological Assay

Even though there are very sophisticated routes to the synthesis and structure proof of PAF and related analogs and homologs, the basic question always will remain whether the experimental sample has biological activity. The possible presence of the 1-*O* fatty acyl analog and other short-chain acyl groups at the *sn*-2 position also must be addressed. The most satisfactory assay to date utilizes the platelet as the test cell. While either the rabbit or the human platelet can serve equally well for the assay, the human platelet has proved to be the cell of choice in the author's laboratory. This decision was based on the ease of collection and the economics of the blood withdrawal process. If the intent is to run PAF bioassays often, then a group of rabbits must be maintained; they are usually kept on a rotation pattern of 20 days after blood collection (by ear vein). The costs of this approach can be sizable. On the other hand, adult humans are readily available and are willing to provide blood at a modest fee. Not only is sufficient blood collected in a short period of time, but over the period of a year, the costs for human blood collection are easily one-tenth that of the rabbit blood process.

A facile method for the isolation of human platelets of high sensitivity to PAF has been described in Chapter 3. This assay is based on the aggregation of the platelets by this agonist. The procedure can detect PAF at concentrations as low as 1×10^{-9} M.

INFLUENCE OF SUBSTITUENTS ON BIOLOGICAL ACTIVITY. The following table illustrates effects of alterations in the sn-1 and sn-2 position of the PAF molecule on its ability to aggregate platelets. The numbers are the EC_{50} values in molarity, M.

	sn-1 Position	
sn-2 Position	1-O-Alkyl Form	1-O-Acyl Form
Acetyl	1×10^{-9}	1×10^{-6}
Proprionyl	1×10^{-8}	1×10^{-6}
Butynoyl	1×10^{-6}	1×10^{-5}
Hexanoyl	1×10^{-5}	No detectable activity

Other modifications to the alkyl form of PAF, such as base-catalyzed methanolysis, leads to an inactive product, namely, 1-O-alkyl-lyso-sn-glycero-3-phosphocholine. In a similar manner, changes in the polar head group lead to decreased biological activity. Examples would be of the removal of the methyl groups (on the choline) leading to a O-phosphoethanolamine derivative that has very low biological activity. A similar result occurs if the phosphate ester bond is changed to a phosphonate (or a phosphinate). However, as shall be described in a later section, removal of the base group and the sn-2 acyl residue can lead to a potent agonist.

Stereochemical Configuration

It is not really possible to determine the stereochemical form of PAF by any of the above procedures. Usually the very low amounts of PAF, coupled with the fact that this compound and its analogs have a very low specific rotation, preclude its measurement by the usual optical means. Perhaps the best analytical tool is phospholipase A$_2$, which was been shown to attack the sn-3 configuration of PAF, for example, readily and to completion. This enzyme has no activity toward the sn-1 form and only cleaves 50% of an sn-2 enantiomer. Neither the sn-1 or sn-2 forms have been found in any PAF isolated from a naturally occurring source.

Experimentally, it is possible to incubate a radiolabeled PAF sample, as described above, with phospholipase A$_2$, and at the end of the desired time period the usual chloroform–methanol extraction can be performed. The chloroform-soluble fraction is applied to a silica gel G thin-layer plate, with appropriate standards in a separate lane, and the plate developed in a solvent system of chloroform–methanol–water (65:35:7, v/v). If the radioactivity is now located at an R_f value near 0.10, which would be comparable to lysophosphatidylcholine, then it would be realistic to conclude that the PAF or its analogs possessed the *sn*-3 configuration.

Synthetic Methods

The chemical synthesis of PAF and related compounds, with the desired stereochemistry, can be achieved as detailed in a review by Hanahan and Kumar (1987). Over the past 10 years the availability of pure, well-defined synthetic PAF of the desired stereochemical configuration has been made possible by several commercial biochemical supply houses. This certainly makes life much easier for investigators in this field.

Structure Proof and Identification of Species

The routes to the proof of structure and identification of species of PAF and related homologs and analogs can be classified into two methodological approaches. The first one can provide limited, but meaningful, data and can be accomplished in any laboratory equipped with basic chemical and biochemical capability, and the second one centers on the use of highly sophisticated (and expensive) instrumentation. These will be referred to simply as Approach I and Approach II, respectively.

APPROACH I.   If the objective of the experimental program is to prove the presence of PAF-like compounds in a biological reaction, usually there will be very small amounts of these products, which are probably insufficient to determine the phosphorus content. However, it is possible to run certain basic assays that will provide some insight into the nature of the material. In the usual instance, experiments are conducted with radioactive precursors (e.g., $^{32}P_i$, [^{3}H]glyceryl ether) in the reaction mixture. Consequently at the end of the reaction period, the lipids are isolated by the usual extraction procedure and subjected to the following assay.

Thin-Layer Chromatography.   As has been described many times before in this book, the sample is spotted on a silica gel G-coated plate, and standard phospholipids (e.g., phosphatidylcholine, PAF, sphingomyelin and lysophosphatidylcholine) are also spotted on a separate lane. The plate is developed in a neutral solvent, chloroform–methanol–water (65:35:6, v/v). Now two possible avenues of exploration are presented and are described as follows:

The first is to cover the sample lane with a glass plate and spray the standard(s) lane with the phosphorus spray. The positive spots are circled carefully (actually dotted would be better) with a thin needle. Then the plate is scanned for radioactivity as described in earlier chapters. If PAF or related derivatives are formed, then radioactivity should be located below the sphingomyelin and above the lysophosphatidylcholine at an R_f value comparable to the standard PAF. This would be presumptive evidence for PAF or a closely related compound.

The second method is based on the results of the first one, and this is to section the sample lane(s) in 0.5-cm sections starting at the origin line and then assay each for biological activity. It is wise to run two to three different volumes of the reaction sample on a plate (in separate lanes, of course) since the amount of biologically active material may be very low. If the area with an R_f comparable to standard PAF contains biologically active material, then this again is presumptive evidence for the presence of PAF-like compounds in the reaction sample.

Another venue would be to subject the unknown sample to base-catalyzed methanolysis (see Chapter 4) and reexamine the location of radioactive material on another thin-layer chromatogram. If the radioactivity is now located at an R_f comparable to standard lysophosphatidylcholine, then this again is support for a PAF-like compound.

Biological Activity. As discussed earlier, it is possible to assay for biological activity through use of rabbit or human platelets and measure the ability of the sample to aggregate these cells (see Chapter 4). This procedure was described earlier, and again a positive result could support the presence of a PAF-like compound(s) in the reaction sample. It is still possible, however, that the biological activity may be low. If a PAF analog—such as long-chain acylacetylglycerophosphocholine, which has low platelet aggregation ability—is present, then base-catalyzed methanolysis should lead to complete loss of radioactivity (if ^{32}P is used) in a chloroform extract of this reaction.

It is important to be aware of the possible presence of an inhibitor in a naturally derived PAF sample (Miwa et al., 1987). Hence a low biological activity in the sample may reflect this fact and/or possibly a high content of the long-chain acyl/short-chain acyl analogs as mentioned earlier.

Chemical Derivatization Technique. A new exciting development in the area of structure proof of PAF is that of chemical derivatization. Satsangi et al. (1989) reported that treatment of PAF with heptafluorobutyric acid anhydride resulted (in a single step) in the formation of the 1-*O*-alkyl-2-acetyl-3-heptafluorobutryoyl derivative in excellent yields. The latter compound could be recovered into chloroform, analyzed by gas-liquid chromatography, and compared with standard samples. Amounts of material as low as 10 pg could be analyzed quantitatively by this procedure. Interestingly, in the course of the original reaction with the anhydride, the polar head group was recovered quantitatively in a water-soluble fraction. Then the conversion of the polar head group to the *t*-butyldimethylsilyl derivative allowed easy identification and quantitation by gas-liquid chromatography. The details of this overall reaction will be discussed further in Approach II.

APPROACH II. This approach relies very heavily on mass spectrometry and as such requires highly sophisticated instrumentation; and, as mentioned be-

fore, the cooperation of a highly talented, knowledgeable mass spectrometrist. As noted earlier, this is an expensive approach, and yet it is such a
powerful and exciting methodology that every effort should be expended to
develop such a resource. In the following discussion, attention will center on
examination of PAF and lysoPAF.

General (PAF and lysoPAF).   The basic objective here is to examine the
spectrum of a PAF sample subjected to fast atom bombardment–mass spectrometry (FAB-MS). The predominant feature of the resulting spectrum is the
formation of a protonated mass ion. In actual fact, there is some cleavage of
the molecule, but it is not extensive. These ions can, however, be used for
diagnostic purposes.

As described in an earlier chapter, the sample is applied to a thioglycerol
matrix on the probe and then subjected to fast atom bombardment. The
spectral pattern for PAF, or 1-*O*-alkyl-2 acetyl-*sn*-glycero-3-phosphocholine,
reveals the following ions: [MH$^+$] protonated mass ion, m/z 524 (16:0 alkyl
side chain); m/z, 552 (18:0 alkyl side chain); m/z 550 (18:1 alkyl side chain);

$$m/z\ 224,\ CH_2{=}CH{-}CH_2O\overset{\displaystyle O}{\overset{\|}{\underset{\underset{\displaystyle O^-}{|}}{P}}}OCH_2\ CH_2N(CH_3)_3;\ \text{and}\ m/z\ 184\ (\textit{O}\text{-phospho-}$$

choline. The latter two ions are produced in small amounts.

Conversion of PAF to lysoPAF, usually by alkaline methanolysis, yields
the following profile: [MH]$^+$, 482(16:0), 510 (18:0), and 508 (18:1) The other
peaks found in this spectrum are similar to those found with PAF. As is
obvious, the differences between the PAF and the lysoPAF pattern is in the
loss of an acetyl group from the PAF.

The limit of detection using this methodology is approximately 10 ng.
Quantitation is not easily or reliably obtained by this technique. However, the
reader is referred to a paper by Clay et al. (1984), in which a deuterated
internal standard is used for estimation of PAF. There are certain problems
with this approach, unfortunately, and these are addressed in a review by
Hanahan and Weintraub (1985). Particular attention is focused on the interaction of the sample with the support matrix, namely, thioglycerol.

In an interesting experimental protocol, Silvestro et al. (1993) utilized
HPLC-mass spectrometry with an ion spray (electrospray) interface for determination of PAF and lysoPAF in human PMN (neutrophils). Both unstimu-
lated and stimulated (with complement-activated zymosan) cells were used as
starting material. The total lipids were isolated in the usual way, and the PAF
was isolated and purified by a combination of thin-layer chromatography,
HPLC, and silica chromatography. This final PAF preparation was subjected
to a bioassay with the inclusion of ^{3}H 16:0 PAF to monitor recoveries.

The PAF-containing samples were applied to a C$_{18}$ reversed-phase column
and for the HPLC separation a mobile phase of methanol–propanol–hexane–
0.1 M aqueous ammonium acetate (100:10:2:5, v/v). The effluent from the

HPLC column was directed into a triple quadropole mass spectrometer equipped with an atmospheric pressure articulated ion spray (electrospray) source. Using MS-MS conditions, dissociation of the parent ions yielded daughter ions comparable to the $[M+H]^+$ of PAF and other molecules, such as O-phosphocholine. Quantitative analyses were obtained by selective ion monitoring. Amounts as low as 0.3 ng of PAF and similar compounds were detected. Interestingly with the lyso derivatives, the limited of detection was 3 ng. The primary alkyl chain lengths present in the neutrophil-derived PAF were 16:0 and 18:0. No mention was made of the presence of any 1-O-long chain acyl analogs.

Chemical Derivatization. Assay techniques for PAF and related compounds that required a derivatization step are not new. For many years, phospholipase C has been a favored reagent to allow generation of a diglyceride from a parent phosphoglyceride sample. The diglycerides could be converted to a form suitable for GC-MS examination. Also a chemical procedure such as acetolysis yields compounds that are satisfactory for analysis. However, there are some serious drawbacks to these approaches, one of which is an incomplete reaction or destruction of sample or product. Hence the recent advent of a facile, direct (single step) chemical derivatization, under mild conditions, has much merit. The seminal publications on this subject were those of Satsangi et al. (1989) and Weintraub et al. (1990). The essential features of this methodology are outlined as follows.

In these investigations, the authors subjected PAF to direct derivatization with pentafluorobenzoyl chloride (PFB) or heptafluorobutyric anhydride (HFB). Using a 16:0 PAF, the resulting reaction sequence is presented in Figure 6-3. The reaction mixture is evaporated under a stream of nitrogen, and the residue is partitioned between hexane and water. The derivatives noted above are found in the hexane layer, and the polar head group is located in water-soluble fraction.

The derivative can be subjected to chemical ionization gas-liquid mass spectrometry. Various naturally occurring homologs (alkyl chain) can be detected by this maneuver. As mentioned before, at the time of the initial reaction with these reagents, there is release of the polar head group which can be recovered in a water-soluble fraction. It can be converted to the t-butyldimethylsilyl derivative and analyzed by FAB-MS. Thus, a complete structure proof of the parent PAF molecule can be realized. If this procedure proves successful in other laboratories, then it will certainly outmode the use of phospholipase C for production (and further analysis) of diglyceride(s) and phosphocholine. The problem with phospholipase C as a structure proof reagent rests on variability in its attack on a substrate, usually depending on the source of this enzyme. In addition, there is possible intramolecular migration of the acyl residues on the diglycerides during the enzymatic reaction. Finally the reaction may not go to completion, and there may not be an equal attack on all species of the PAF or other substrate. Stimulation of human leucocytes

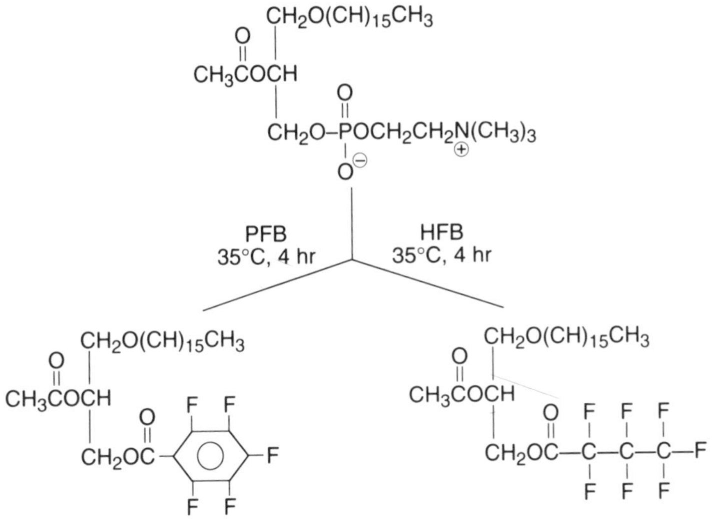

FIGURE 6-3. Direct derivatization of platelet activating factor for structural proof analysis.

allowed detection of 16:0, 17:0, 18:0, and 18:1 PAF by the above anhydride reaction at levels as low as 50 pg.

In a particularly definitive investigation, Balazy et al. (1991) explored the assay of PAF and alkylether phosphoglycerides, subsequent to derivatization, by gas-liquid chromatography coupled with mass spectrometry. Both commercial PAF and naturally occurring PAF (from neutrophils stimulated with calcium ionophore) were studied. These compounds were derivatized to the pentafluorobenzoyl form and analyzed by a combination of electron capture, gas–liquid chromatography, and mass spectrometry coupled with a stable isotope assay. Excellent detail is provided on the procedures used, the recovery of PAF from the cells, the derivatization step, and the subsequent analyses. Levels of PAF at 2–3 pmol/10^6 neutrophils could be determined by this methodology. Standard curves (using well-defined PAF) showed a linear response in the range from 5 to 50 pg.

PAF Analogues

Although the occurrence of analogs of PAF in mammalian cells was always considered a possibility, Mueller et al. (1984) first reported the presence of 1-O long-chain acyl-2-acetyl-glycero-3-phosphocholine in rabbit neutrophils. Later, Tokumura et al. (1989) isolated a vasopressor active phosphatidylcholine fraction from bovine brain. Treatment of the fraction with phospholipase C yielded a diglyceride component which was converted to the t-butyldimethylsilyl derivative. Analysis of the latter by gas-liquid chromatography coupled with mass spectrometry showed the presence of three types of PAF (hexadecyl, octadecyl, and octadecenyl alkyl chains) and 17 1-O-long-chain acyl analogs. The latter was shown to contain, in the main, the acetyl, propionyl, acryloyl, butyroyl, valeroyl, caproyl, and heptanoyl forms. Though these 1-O-acyl species were present in significant amounts, they exhibited minimal depressor activity compared to the 1-O alkyl analogs.

Recently, Weintraub et al. (1991) reported that the 1-O-acyl analogs of PAF could be determined by the chemical derivatization method described earlier. Using well-defined 1-O-acyl standards, they showed that the heptafluorobutyric anhydride technique allowed formation in excellent yields of the heptafluorobutyroyl derivative. The latter could be analyzed by gas-liquid chromatography coupled with mass spectrometry (electron capture). The limits of detection were in the picogram range. It will be interesting to see how well this approach works on mixtures of naturally occurring 1-O-alkyl and 1-O-long-chain acyl phosphoglycerides.

Lysophosphatidic Acid (lysoPA) and Phosphatidic Acid (PA) and Their Ether Analogs

It has been established for many years that lyso(acyl)PA and (diacyl)PA are important intermediates in the biosynthesis of phosphoglycerides. As discussed in a mini-review by Exton (1990), agonist signaling also can lead to the formation of phosphatidic acid. Though not mentioned in this latter review, it is entirely possible that lysoPA also can be formed. Interestingly, though, over a period of several years the occurrence and biological activity of lysophosphatidic acids have been recognized and explored to a considerable degree. A review by Moolenaar (1994) summarizes the available information on the hormone-like and growth-factor-like activities of the lyso(acyl)phosphatidic acids.

While primary experimental attention has been directed toward the monoacylglycerophosphoric acids and the diacylglycerophosphoric acids, it has become increasingly evident that their ether analogs can occur in biological systems. Their chemical structures are provided in Figure 6-4, where R_1 and R_2 represent long-chain hydrocarbon chains. Inasmuch as these lipid phosphoric acids can be derived from their corresponding diacyl- and monoalkylmonoacylphosphoglycerides, it is also possible that an alkenyl(vinyl ether) analog is formed, but to date it has not been detected or studied.

These lipid phosphoric acids possess a spectrum of biological activities of importance. The ether analogs, in particular, exhibit agonist activity at levels comparable to PAF. Unfortunately these compounds are found in very small amounts in biological tissues. As expected, this fact compromises their analytical evaluation. Nevertheless, there are certain procedures that can be undertaken to establish, with some certainty, their structural characteristics in naturally occurring samples. These approaches are discussed in the following sections.

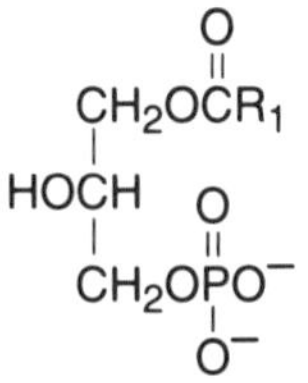

1–O–long-chain acyl
glycerophosphoric acid
[lyso (acyl) PA]

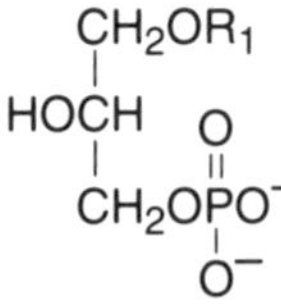

1–O–long-chain alkyl ether
glycerophosphoric acid
[lyso (alkyl) PA]

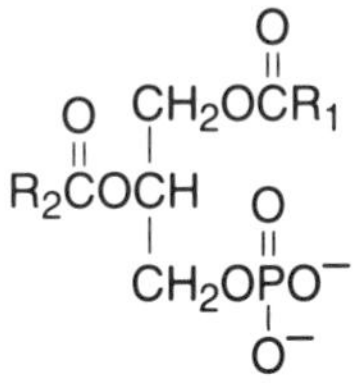

1–O–long-chain acyl
2–O–long-chain acyl
glycerophosphoric acid
[diacyl PA]

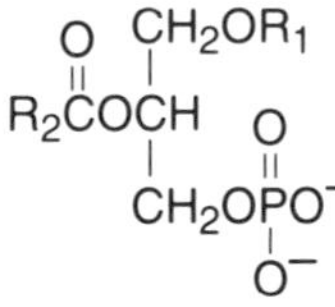

1–O–long-chain alkyl ether
2–O–long-chain acyl
glycerophosphoric acid
[alkyl (acyl) PA]

FIGURE 6-4. Chemical structure of naturally occurring glycerol containing (lipid) phosphoric acids.

Isolation and Purification

Though the lipid phosphoric acids are present in very low concentrations even in stimulated cells, it is possible with certain precautions to isolate them in good yields. Perhaps the most important experimental factor in a successful extraction is the inclusion of acid, usually 1 N HCl, in the extracting solvent. This is important because these acidic compounds normally exist as salts in the cell; in order to extract them into a chloroform-rich fraction, for example, they must be converted to the free acid form. Given the satisfactory isolation of the total lipids, aliquots of this fraction can be assayed for total phosphorus by the "macro" procedure described in Chapter 4 or by the submicrogram method described by Böttcher et al. (1961). Two approaches can be used to separate the lysoPAs and the PAs from the other lipids as outlined on the following section.

Thin-Layer Chromatography

As noted above, depending on the amounts of material available and the objectives of the experimental protocols, two chromatographic procedures can be undertaken and are described as follows.

ONE-DIMENSIONAL. It is assumed at this point that the major objective of the study is to isolate and structurally identify the lipid phosphoric acids. A particularly effective separation is that described by Mauco et al. (1978). A total lipid sample is applied to a silica gel G plate impregnated with 0.25 M oxalic acid. Using a solvent system of chloroform–methanol–12MCl (87:13:0.5, v/v) which was first suggested by Cohen and Derksen (1969), the lipid phosphoric acids can be separated from the majority of the other lipids, which remain at or near the origin. The R_f value for the PAs was near 0.60 and for the lysoPAs was near 0.30. These compounds can be detected using the TNS spray or radioactive monitoring (assuming a radiolabeled precursor were used) (see Chapter 4). The desired areas can be recovered from the adsorbent with a solvent system of chloroform–methanol–12 M HCl (65:35:5, v/v). Phasing of the latter extract by the addition of water will yield the lipid phosphoric acids in the chloroform-rich phase. The latter extract is washed with methanol–water (10:9, v/v) until the aqueous wash fraction is neutral. The chloroform-rich fraction can be analyzed further, using the following process.

This general procedure is suitable for preparative purposes. Yet it is important to note that acidic compounds, other than the lysoPAs and PAs, can be present. These would be compounds such as cardiolipin and phosphatidylglycerol. The further identification of the compounds isolated by this methodology can be accomplished by two dimensional chromatography as dicussed next.

TWO-DIMENSIONAL.   Actually two approaches can be undertaken at this point. The first is to study the "lipid phosphoric acids," isolated by one-dimensional thin-layer chromatography on Whatman K6 or Merck silica gel 60 plates. A separate plate containing well-defined standards should be run at the same time. The second approach is to use an aliquot of the total lipid sample for separation. In each case, the solvent system of choice for the first direction is chloroform–methanol–ammonia (28%) (65:35:5, v/v), and the second direction is a solvent mixture composed of chloroform–acetone–methanol–acetic acid–water (45:20:10:13:5, v/v). As suggested above, the individual components can be located by the TNS spray. Under these conditions the lysoPAs and the PAs are well separated. Depending on the experimental objectives, the desired areas can be isolated by solvent extraction as described above and assayed for radioactivity, if a radiolabeled precursor was incubated with the cells under study and/or for total phosphorus. Preparative two-dimensional (as well as one-dimensional) thin-layer chromatography can be used for recovery of lysoPAs and PAs and analyzed further as discussed in the following section.

Structural Analyses

Lysophosphatidic Acids (LPA)

It is assumed at this point that a highly purified sample of lysophosphatidic acids has been isolated by thin-layer chromatography. The next step then is to ascertain whether this preparation is a single component, such as 1-*O*-acylglycerophosphoric acid or a mixture of the 1-*O*-acyl-, 1-*O*-alkyl-, and 1-*O*-alkenylglycerophosphoric acids. Although no evidence for the presence of an alkenyl analog has been provided, it is still important to assay for its possible presence. Many of the same techniques described in chapters 4 and 5 for the analysis of the diacyl-, alkylacyl-, and alkenylacylphosphoglycerides can be applied here and are briefly discussed below. Primary attention will focus on the use of acid or base treatment as an initial approach to proof of composition and structure.

BASE-CATALYZED METHANOLYSIS. Use of 0.5 M NaOH in methanol at room temperature for 20 min will result in cleavage of the acyl bond in the 1-*O*-acylglycerophosphoric acid. Thus, if all three species mentioned above are present, examination of the base-treated sample should show two TNS-positive spots: One of these will be at an R_f near 0.85 and should be the methyl ester of the long acyl group, and the other slower moving spot (in a polar solvent system) will be a mixture of the 1-*O*-alkyl and 1-alkenyl glycerophosphophoric acids. If sufficient material is present, it should be possible to detect the alkenyl group by the Schiff base reaction.

This (base) procedure will provide evidence for the presence of a fatty acyl-containing glycerophosphoric acid. In addition, an estimation of the relative amount of this component in the sample can be gained by assay of loss of phosphorus from a chloroform-rich extract of the reaction mixture and the gain of phosphorus (phosphate) in the aqueous fraction.

The preceding information would support the presence of an ester bond in the starting material; if so, then a methyl ester component should be detected by thin-layer chromatography. Subsequent analysis by GC-MS should confirm its presence and structural characteristics.

ACID TREATMENT. In a manner similar to that outlined in Chapter 4, the presence of an alkenyl group can be detected with reasonable ease. The resulting reaction mixture should contain a long-chain fatty aldehyde and unchanged 1-*O*-fatty acyl and 1-*O*-alkyl GPAs. If sufficient material is available, a phosphorus assay on the chloroform-soluble products would show a loss of phosphorus compared to the starting level. This would provide an estimation of the amount of 1-*O*-alkenyl GPA. The long-chain fatty aldehyde liberated in this reaction can be recovered by thin-layer chromatography, converted to the dimethylacetal derivative, and examined by GC-MS.

Using the above methodology, it is possible to obtain the composition of a naturally occurring lysophosphatidic acid(s) preparation. At the present time, the three chemical forms described—the 1-*O*-acyl-, the 1-*O*-alkyl-, and the 1-*O*-alkenylglycerophosphoric acids—are the most likely types to be encountered in mammalian cells.

The further structural characterization of the lysoPAs can be accomplished through use of the techniques described in chapters 4 and 5. Perhaps one of the more effective assays is that of FAB-MS. The type of information likely to be gained by this technique is described in the following section.

FAB-MS SPECTRAL PATTERN.   This potent technique is most effective as an aid in structure proof of lipid (glycerol) phosphoric acids. These compounds produce a protonated molecule [MH]$^+$ as well as several other ions of importance. Analysis of an alkyl (16:0) lysoGPA will show an ion at m/z 397, which is attributed to the [MH]$^+$. If an acyl (16:0) lysoGPA is examined by FAB-MS, an ion is found at m/z 411, which is indicative of its [MH]$^+$. Another important ion is [MH-H$_3$PO$_4$]$^+$, m/z 313 (acyl (16:0) lysoGPA), and m/z, 299 (alkyl 16:0) lysoGPA). Further information on the FAB-MS patterns of these and other phosphoric acid derivatives is summarized in a recent publication by Sugiura et al. (1994).

Even though a FAB-MS pattern for a pure lysoalkenyl GPA has not been reported, it would most likely show an ion at m/z, 395, which would be the [MH]$^+$ for a 16:0 species. However, it is important again to stress that this type of compound would undoubtedly interact with the thioglycerol matrix and produce a rather complex spectrum.

Perhaps the best route to evaluate the composition of a lysoPA mixture is through the methods cited earlier, especially FAB-MS. A subtractive type of analysis can done in which the FAB-MS spectrum of the original sample is obtained; then, after each treatment (acid, base, etc.) the sample is purified by thin-layer chromatography, the remaining lysoPA recovered and examined again by FAB-MS. These various profiles will yield information of value in determination of the structural characteristics of the sample under study. Thus, if after an acid reaction followed by a base treatment a chloroform-soluble phosphorus compound remains, it would most likely be the lysoalkyl ether GPA. The FAB-MS spectrum would support such a conclusion.

STEREOCHEMICAL CONFIGURATION.   Little attention has been paid to the optical conformation of any lysophosphatidic acids found in mammalian cells. This is entirely understandable due to the very small amounts of these compounds normally isolated. However, on the basis of the fact that these compounds are formed in the *de novo* pathway for biosynthesis of phospho-glycerides and that they could arise also by action of an endogenous phospho-lipase A$_2$ action, it is highly probable that they are of the *sn*-3 configuration.

The importance of a specific stereochemical conformation to the lysoPAs active as agonists still is open to debate. For example, an analog of 1-*O*-alkyl GPA, namely 1-*O*-hexadecylpropanediol PA, which has no optical center, has a biological activity that is nearly the same as that of the 1-*O*-alkyl GPA (Sugiura et al., 1994). Inasmuch as only the *sn*-3 form of the alkyl GPA was available for testing in this latter study, it will be of importance to assay the activity of an *sn*-1 enantiomer in the same biological assay system. Then a more meaningful conclusion can be reached as to the importance of a specific optical configuration in evoking agonist activity by the lysoGPAs. The lack of a requirement for a specific stereochemical form is unique.

Phosphatidic Acids

These compounds are well established as important intermediates in the biosynthesis of phosphoglycerides, such as phosphatidylcholine. The fact that they are formed on agonist stimulation of a cell suggested a role in the cell signaling process (Exton, 1990). In addition, over a period of many years, there have been reports on the interrelation of phosphatidic acid formation and cellular responses (Ca^{2+} movements to agonists, for example). Thus, these compounds have emerged as important components of cellular metabolism and merit serious attention in any metabolic investigation.

The phosphatidic acids could potentially be found as a mixture of the 1-*O*-diacyl-, alkylacyl-, and alkenylacylglycerophosphoric acids. Structural characterization follows almost exactly the route as described for the lysophosphatidic acids.

ISOLATION AND PURIFICATION. The methodology described in the section entitled "Lysophosphatidic Acid (lysoPA) and Phosphatidic Acid (PA) and their Ether Analogues" showed that the lysophosphatidic acids and phosphatidic acids can be isolated well separated from each other by thin-layer chromatography. Consequently, preparative thin-layer chromatography can be employed to isolate the phosphatidic acids in sufficient amounts for structure proof studies.

STRUCTURE PROOF. Again there is the question of whether the sample is a single structural type (i.e., a diacyl GPA) or whether it is a mixture containing, in addition, the alkylacyl GPA and alkenylacyl GPA. Most often investigators are not interested in this possibility, have very small amounts of material available for analysis, and/or are unaware of the potential for mixed species. If it is desired to establish with certainty the composition of a phosphatidic acid preparation, exactly the same experimental design as was applied to the lysophosphatidic acids can be used. Essentially this would involve the following.

Base Treatment. Under the same conditions as described above, this would lead to complete degradation of the diacyl GPA with formation of methyl esters and glycerophosphoric acid. Concomitantly there would be a decrease in chloroform-soluble lipid phosphate due to preference of the glycerophosphoric acid for a water-rich phase. In addition, if the alkylacyl and alkenylacyl GPAs are present, there would be formation of lysoalkyl GPA and lysoalkenyl GPA (and methyl esters from the parent sample), and these could be detected on thin-layer chromatography. In the latter case, standards could be used as well as a sample of untreated starting material for comparative purposes.

Acid Treatment. This procedure, if applied to a sample containing all three species, would lead to release of a long-chain fatty aldehyde which could be isolated and analyzed as the dimethylacetal. Subsequent thin-layer chromatography of the acid reaction mixture would allow recovery of the parent diacyl GPA and alkylacyl GPA. Then, using the chloroform soluble lipid phosphate as a guideline, base treatment of the original sample will lead to loss of chloroform-soluble phosphorus (representative of the diacyl GPA). The remaining chloroform-soluble phosphorus will represent the alkyl and alkenyl form.

Thus an estimation of the relative amounts of the diacyl, alkylacyl, and alkenyl acyl GPAs can be obtained through use of the acid and base reactions.

FAB-MS Profile. As with the lysophosphatidic acids, this type of mass spectrometry can be useful in establishing the structural characteristics of PAs in a mixture. Interpretation of the spectral patterns is slightly more complex than for the lysophosphatidic acids due to the presence of additional fatty acyl residues. Nevertheless, a "subtractive" approach as described for the lysophosphatidic acids can be of significant help. Even with the complexities encountered in a sample, the FAB-MS methodology is a potent aid in elucidating the structure of these compounds.

Stereochemical Configuration. Though not a frequently used procedure for establishing the optical conformation of naturally occurring phosphatidic acids, the evidence all suggests that they are of the *sn*-3 configuration. In the instances where the isolated phosphatidic acids were treated with phospholipase A$_2$, it was obvious that they were degraded to lyso compounds in good yield. These results would support an *sn*-3 configuration for these lipid phosphoric acids.

Some General Comments. Although a number of chemical methods for the synthesis of various phosphatidic acids have been published, these procedures are often of low yield and require a deft hand for organic chemical methodology. However, there are several commercial suppliers who have available pure, well-defined phosphatidic acids for research investigations.

Phosphatidylglycerol

Some 40 years ago in the halcyon days of study of the then new technique of silicic acid column chromatography of lipids, it was apparent to several investigators that there were nitrogen-free phospholipids in these samples. Also, these unique, acidic phospholipids did not contain inositol. In the usual protocol, lipids were applied to a silicic acid column in a nonpolar solvent such as hexane (or petroleum ether, b.p. 30–60°C), and elution continued with increasing amounts of diethyl ether. These solvent mixtures would elute the

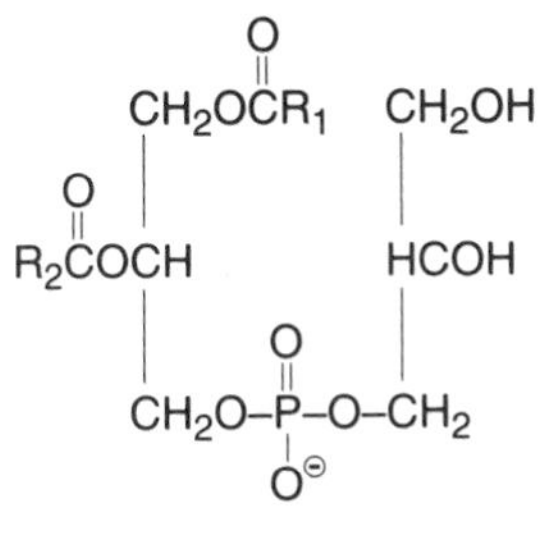

Phosphatidylglycerol
(sn-1,2-Diacyl-sn-glycero-3-phospho-1'-sn-glycerol)

FIGURE 6-5. Structural formula for phosphatidylglycerol.

"neutral" lipids (e.g., cholesterol, cholesterol ester, triacylglycerol, free fatty acids). Subsequent addition of more polar solvent mixtures—for example, chloroform–methanol (98:2, v/v)—would elute the nitrogen-free phospholipids. These lipids were shown to a mixture of phosphatidic acid, cardiolipin, and phosphatidylglycerol. The structure of the phosphatidylglycerol molecule was proven to be as depicted in Figure 6-5. Chemically it is defined as 1,2-diacyl-sn-glycero-3-phospho-1'-sn-glycerol. It is found in mammalian cells, mainly in the mitochondria, but at very low levels, approximately 2% of the total phospholipids. It is in high concentrations (40–60% of the total phospholipids) in plant chloroplasts and in variable amounts in bacteria. A dialkyl form was shown to be present in halophilic bacteria by Joo and Kates (1969). However, in mammalian cells no evidence has been presented for the dialkyl form or for the more conventional alkyl or alkenyl analogs.

Isolation, Purification and General Identification

Isolation

The most effective procedures for isolation of phosphatidylglycerol from a cell take advantage of methodology such as thin-layer chromatography, silicic acid chromatography, and high-pressure liquid chromatography (HPLC). In all of these techniques, the major obstacle, if the investigation centers on mammalian cells, is the small amount of phosphatidylglycerol present. This can be circumvented, in part, by the use of radiolabeled precursors in cellular experiments and by the use of standards, radiolabeled or not, as chromatographic markers. Usually the latter are run on separate lanes in thin-layer chromatography and as a marker on column chromatography.

The most widely used solvent extraction technique for recovery of phosphatidylglycerol from a cell or cell preparation is through use of chloroform–methanol mixtures. It is advisable to include acid because this and similar compounds are acidic in nature and are associated often with cations in the cell. The phosphatidylglycerol partitions very nicely into the chloroform-rich extract and is quite stable for a period of time if stored in a chloroform–methanol mixture containing a low amount of the antioxidant BHT.

Purification and General Identification

THIN-LAYER CHROMATOGRAPHY.   This can be accomplished on silica gel G (or HP60) plates (250 μm) using a solvent system of chloroform–methanol–ammonia (28%, v/v). If sufficient phosphatidylglycerol is present, it can be located by phosphorus spray, sulfuric acid char, TNS spray, or the periodate–Schiff base spray. In the usual instance, phosphatidylglycerol would be found at an R_f value near 0.37, with phosphatidylcholine at R_f 0.25 and phosphatidylethanolamine at R_f 0.48. As noted earlier, a separate lane containing pure standards can be most helpful in identifying individual phospholipid classes.

If desired, preparative thin-layer chromatography can be accomplished by using 500- to 1000-μm plates on which as much as 10 mg of total lipid can be separated in the same solvent system as described previously. Then, the desired area can be removed by scraping and extracting with acidified chloroform–methanol (2:1, v/v). Subsequent phasing of this extract and washing of the chloroform-rich phase with methanol–water (10:9. v/v) will give the presumed phosphatidylglycerol.

It is important to note that the thin-layer chromatographic patterns of phosphatidylglycerol, as well as the other acidic phosphoglycerides (this could include phosphatidic acid, cardiolipin, phosphatidylserine), are influenced significantly by the adsorbent and the solvent system. For example, if these compounds are applied as the sodium salts on a silica gel G plate and the chromatogram is developed in a neutral solvent, chloroform–methanol–water (65:35:7, v/v), there is sufficient calcium ion in this adsorbent to cause an ion exchange effect and the final spot on the plate will represent the calcium salt of the lipid. If the sample is run in a solvent system of chloroform–methanol–ammonium hydroxide (65:35:7, v/v), the acidic lipid will migrate as the ammonium salt. Thus, the use of highly purified standards for comparison of R_f values is highly recommended.

SILICIC ACID COLUMN CHROMATOGRAPHY.   Perhaps one of the more successful routes to isolation of phosphatidylglycerol from a total lipid extract is by silicic acid column chromatography; columns containing as little as 1 g of silicic acid (SilicAR CC-7, Merck) have been used with success. As discussed in an earlier chapter, there should be a 10:1 height-to-diameter ratio for the column bed. Subsequent to application of the sample, either hexane–diethyl ether (1:10, v/v) or pure chloroform can be used to elute the neutral lipids. The elution can be followed by spotting aliquots of the eluent on micro-silica gel plates and testing for char and phosphorus reaction. Then careful application of chloroform–methanol in ratios of 98:2, (v/v) to 90:10 (v/v) will elute an enriched phosphatidylglycerol fraction. If sufficient material is available, repeated chromatographic separation on fresh columns can be done under the same conditions as above. The phosphatidylglycerol-enriched fractions can be combined and rechromatographed on a new silicic acid column. Purity of the sample can be evaluated by analytical thin-layer chromatography (250-μm plates) using three different solvent systems: neutral, acidic, and basic.

HIGH-PRESSURE LIQUID CHROMATOGRAPHY (HPLC).   This technique is a sophisticated, but very helpful, route to the separation of individual classes of phospholipids, including phosphatidylglycerol. Recently, Bernhard et al. (1994) described an isocratic HPLC system for fractionation of individual classes of phospholipids. Using a Nucleosil 5 NH$_2$ column and a mobile phase of acetonitrile–methanol–water–methylphosphonic acid (50% in water), 1460:500:30:0.6, v/v), separation of a wide variety of phospholipids could be achieved within 60 min. Phosphatidylglycerol was found to elute after lysophosphatidylcholine, sphringomyelin, and phosphatidylcholine and immediately (3–5 min) before phosphatidylethanolamine. Detection was accomplished by measurement of the ultraviolet absorption of the eluent at 205 nm and also by a post-column fluorescence assay. Often the phospholipids containing the more saturated hydrocarbon moieties fail to be detected (at 205 nm), and the fluorescence assay using 1,6-diphenyl-1,3,5-hexatriene added a new dimension to the detection process. The latter reagent formed micelles with the phospholipids, and their presence was followed by excitation at 340 nm and emission at 460 nm. In this procedure, the degree of saturation of the individual phospholipids did not affect the fluorescence, and the sensitivity was essentially the same for all classes. The lower limit of detection was close to 0.5 nmol.

Structure Proof

The methods to be outlined here can be used, with careful choice of conditions, to establish with certainty the structural characteristics of a presumed phosphatidylglycerol sample. In most studies, at least with mammalian cells, the amount of material available for assay will be small; hence thin-layer chromatographic evaluation of the success of a procedure will be an invaluable aid.

Base-Catalyzed Methanolysis

Incubation of a sample of phosphatidylglycerol with 0.5 M KOH in methanol at room temperature for 15 min will lead to complete cleavage of the fatty acid ester bonds with formation of methyl esters of long-chain fatty acids and *sn*-glycero-3-phospho-1′-*sn*-glycerol. The progress of the reaction can be followed by loss of chloroform-soluble lipid phosphorus and the appearance of the methyl esters. This derivative can be further characterized by GC-MS. Examination of the water-soluble products will show that over 95% of the starting phosphorus is recovered. Periodate uptake should show a *vic*-glycerol/P molar ratio of 2.0, compared to a value of 1.0 for the starting preparation.

Phospholipase C Cleavage

Phosphatidylglycerol can be hydrolyzed readily at room temperature by this enzyme using the methodology described in Chapter 4. The cleavage products will be *sn*-1,2-diacylglycerol and *sn*-1-glycerophosphoric acid. The diacyl derivative is extracted into chloroform, while the glycerophosphoric acid remains in the water-soluble fraction. The diacylglycerol can be purified by thin-layer chromatography, and the desired area can be scraped and extracted with chloroform–methanol (2:1, v/v) and converted to the *t*-butyldimethyl-chlorosilyl derivative and subjected to GC-MS as described in Chapter 4.

The water-soluble, phosphorus containing fraction will consume 1 mol of periodic acid per mole of phosphorus if it is the *sn*-1-glycerophosphoric acid. If this is indeed the *sn*-1 conformer, then it should not react with glycero-3-phosphate dehydrogenase. A control with the *sn*-3 form will show nearly a 100% reaction. This result would be supportive of the *sn*-1 configuration for the "free" glycerol portion of the molecule.

Phospholipase D Attack

This enzyme can catalyze formation of phosphatidic acid and free glycerol. The progress of the reaction can be followed by thin-layer chromatography. A standard sample of phosphatidic acid should be run in an adjoining lane for comparison purposes.

If further information is desired on the structure of the compound assumed to be phosphatidic acid, then it can be extracted from a preparative plate and subjected to base-catalyzed methanolysis. The products should be the methyl esters of the long-chain fatty acids (from the parent compound) and *sn*-glycero-3-phosphoric acid. The latter product, which will be water-soluble, is assayed for total phosphorus, *vic*-glycerol, and *sn*-3 glycerophosphoric acid content (with glycero-3-phosphate dehydrogenase). The result should show a *vic*-glycol/P molar ratio of 1.0 and a *sn*-3 glycerophosphoric acid/P molar ratio of 1.0. Controls should be run with a standard *sn*-3-glycerophosphoric acid and an *sn*-1-glycerophosphoric acid, if available.

Phospholipase A₂ Cleavage

Phospholipase A_2 (snake venom) will attack phosphatidylglycerol smoothly and to completion under conditions previously described in chapter 4. One mole of fatty acid is released per mole of lipid phosphorus. This result supports an *sn*-3 configuration for the phosphatidic acid portion of the molecule.

Diacylglycerol Release by Periodic Acid/ Dimethylhydrazine Treatment

A strictly chemical approach to liberation of the diacylglycerol backbone of phosphatidylglycerol was reported by Heinze et al. (1984). In this procedure, the sample is reacted with periodic acid in methanol in the dark for 90 min at room temperature. Chloroform and 0.45% NaCl are added with vigorous mixing, and the chloroform-soluble material was recovered. This will contain the peroxidized lipid (1 mol of formaldehyde liberated per mole of phosphorus), which is subsequently treated with 1,1-dimethylhydrazine. The final product is phosphatidic acid, which can be isolated by thin-layer chromatography as described earlier and further studied, if desired. This technique would give further support to a phosphodiester linkage between the diacyl glycerol and the "free" glycerol portions of the phosphatidylglycerol molecule.

COMMENTS.   This structure proof approach should provide a reasonable basis for reaching a conclusion as to the structure and structural characteristics of a phosphatidylglycerol preparation. Some of experimental approaches described above were taken in part from the classical observations by Haverkate and van Deenen (1964, 1965) and by Gray (1964).

The proof of structure of the nitrogen-free, acidic phosphoglycerides requires close attention to their purity. Their close physical characteristics, especially their acidic charge, can be a problem due to the interaction of these groups with cations in the chromatographic adsorbents (and reagents). Thus, R_f values alone should not be the only criteria used in identification of a particular compound. The other methodologies described in this chapter for structure proof (and also elsewhere in this book) can be of invaluable assistance in the characterization process.

Cardiolipin

Cardiolipin, also referred to as diphosphatidylglycerol, is a unique phosphoglyceride found almost exclusively in the mitochondrial membranes of eukaryotic cells. It can account for as much as 20% of the mitochondrial membrane lipids. The chemical structure of this compound has been well established and is shown in Figure 6-6.

Chemically it can be defined as 1,3-bis(sn-3-phosphatidyl)-sn-glycerol. A particularly fascinating feature of this molecule is its fatty acid composition. In many preparations, linoleic acid (18:2) can represent as much as 90 mol %, with the remainder being composed largely of oleic acid (18:1) and palmitoleic acid (16:1). It is not uncommon for cardiolipin samples to contain 95–98 mol % unsaturated fatty acids, with linoleic acid being the major constituent.

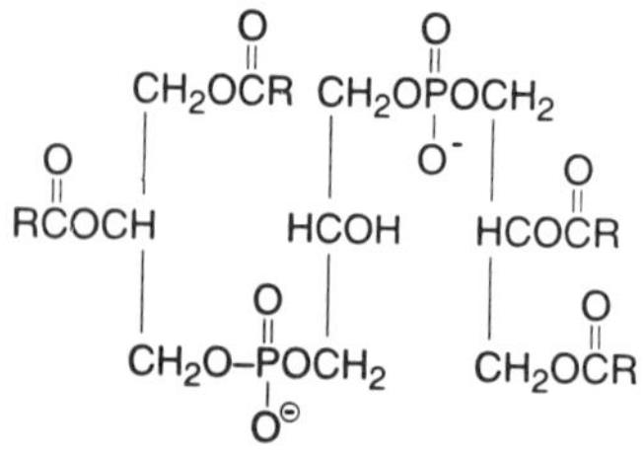

$$\underset{\substack{| \\ O^{\ominus}}}{\overset{O}{\underset{}{\parallel}}}$$

Cardiolipin
(Diphosphatidylglycerol)

FIGURE 6-6. Chemical structure of cardiolipin (diphosphatidylglycerol).

As might be expected, such an unusual compound has been the subject of intense biochemical study over the years. It is closely associated with the function of the cytochrome oxidase in the inner mitochondrial electron system. An introduction to this field of endeavor can be gained from publications by Robinson et al. (1990) and by Hoch (1992).

Isolation, Purification, and General Identification

Cardiolipin can be isolated in sufficient quantity for structure proof studies from a tissue such as bovine heart, where gram quantities of total lipid can be obtained. These techniques then can be applied, with care, to experimental protocols in which nanogram levels of total lipids are encountered.

Isolation

A mixture of chloroform–methanol (acidified with 0.5 N HCl), as described in earlier sections, will allow an excellent recovery of this phosphoglyceride from cellular preparations. Subsequently, the chloroform-rich phase is washed with methanol–water (10:9, v/v) until acid-free. This total lipid extract then can be subjected to the following procedures.

Purification and General Identification

SILICIC ACID CHROMATOGRAPHY.   This is an ideal preparative route to isolation of a cardiolipin-rich fraction. A favorite column packing is SilicAR, CC-7 (Mallinckrodt). Application of the total lipid sample in petroleum ether (b.p. 30–60°C) and subsequent elution with petroleum ether–diethyl ether will remove the neutral lipids (cholesterol, free fatty acid, triacylglycerol, cholesterol esters), if a fractionation of the neutral lipids is desired. Otherwise, application of the total lipid sample in chloroform will elute the neutral lipids together. Then application of mixtures of chloroform–methanol (v/v ranging from 100:1 to 95:1) will elute the nitrogen-free acidic phospholipids represented by phosphatidic acid, phosphatidylglycerol and cardiolipin. In the usual instance, chloroform–methanol (50–60:1, v/v) will elute phosphatidic acid, chloroform–methanol (19:1, v/v) will elute cardiolipin and finally chloroform–methanol (12:1, v/v) will elute phosphatidylglycerol. Often these fractions do not emerge in pristine purity, but often are contaminated with pigment or some of the other acidic phosphoglycerides. Thus, repeated silicic acid column chromatography and/or thin-layer chromatography can resolve this problem.

Smaal et al. (1985) described a novel procedure for the purification of cardiolipin, as the sodium salt, from beef heart. Extraction of the total lipids was achieved by the usual mixture of chloroform–methanol, followed by acetone precipitation of the phospholipids and conversion of the acidic phosphoglycerides to the calcium salts. From the latter, a cardiolipin-rich fraction was obtained by partition column chromatography on silica gel. A purification step involved HPLC of the calcium salts on silica gel and their conversion to the sodium salt form. This material was subjected to HPTLC (high-performance thin-layer chromatography) in the solvent system of chloroform–methanol–acetic acid–water (90:40:12:2, v/v) and showed a single, concise spot at an R_f value near 0.80. The sodium salt of cardiolipin was dissolved in benzene and stored under argon, at $-80°C$ in sealed glass vials. Analysis of this product showed it to have a fatty acid/P molar ratio of 2.03, a sodium/P molar ratio of 0.99, and 90.7 mol % linoleic acid (a total of 98 mol % unsaturated fatty acid). This fatty acid profile was considered to be diagnostic for cardiolipin. However, it would have been helpful to have further analytical proof for this structure. A fatty acid/P molar ratio of 2.0 and a sodium/P molar ratio of 1.0 could apply to several acidic lipids. Of particular help would have been a mass spectrum on the purified material.

THIN-LAYER CHROMATOGRAPHY.   Two-dimensional thin-layer chromatography is often the preferred route to analytical evaluation of the presence of cardiolipin in a biological sample or further purification of cardiolipin obtained by silicic acid chromatography. Using silica gel containing alkaline magnesium silicate, Hovius et al. (1990) reported a good separation of the phospholipids of liver mitochondria. A spot attributed to cardiolipin was well separated from the other phosphoglycerides.

One-dimensional thin-layer chromatography has been used to a limited extent. It will not cleanly separate phosphatidic acid, phosphatidylglycerol, and cardiolipin from each other in a neutral solvent system. However, it can be of value in evaluating the purity of cardiolipin-rich fractions as demonstrated by Smaal et al. (1985).

SILICIC ACID HPLC.   Cardiolipin and certain of its derivatives, namely, monolysocardiolipin (one fatty acid removed) and dilysocardiolipin (two fatty acids removed), can be separated by this technique. Amounts as low as 40 nmol could be detected and analyzed quite smoothly by silicic acid HPLC as described by Robinson (1990). This novel separation procedure, using hexane–2-propanol–1 mM phosphoric acid (50:50:3.5, v/v) as the solvent system and monitoring of the eluent at 203 nm, can separate quite cleanly cardiolipin, lysocardiolipin, and dilysocardiolipin.

Structure Proof

Assuming that a highly purified sample of cardiolipin is now available, the following reactions can be used to prove its structure.

Base-Catalyzed Methanolysis

This procedure can provide significant information, using exactly the same conditions as described earlier. A sample is dissolved in a small volume of chloroform, and 10 volumes of 0.5 N KOH in methanol is added and the mixture is then incubated for 20–25 min at room temperature. This reaction mixture is neutralized with 6 N HCl, and sufficient chloroform is added to allow phase separation. Following vigorous mixing, the chloroform-rich lower fraction and the water-rich upper fraction are recovered, the chloroform-rich fraction, which should be free of lipid phosphorus, can be analyzed by GC-MS, and the fatty acid composition can be determined. The glycerol-containing backbone of the original molecule is recovered in the water-rich fraction. The total phosphorus content of this fraction should agree very closely with that of the original starting material. Further analysis of this extract is discussed in the following section.

Periodate Cleavage

The water-soluble fraction can be subjected to periodic acid oxidation as illustrated in the reaction sequence given in Figure 6-7.

The fatty acid free derivative, II, is usually optically active and strongly suggests that the two glycerophosphate residues have the same stereochemical configuration. Otherwise this derivative would have had a *meso* configuration (with no optical activity). The peroxidation of II will yield 2 mol of formaldehyde per mole of phosphorus together with a dialdehyde, III, which on reaction with dimethylhydrazine will yield glycerol-1,3-diphosphoric acid, IV.

These results are supportive of the structural formula for cardiolipin presented earlier in this section.

Phospholipase A_2 Attack

Phospholipase A_2 isolated from snake venom *(Crotalus adamanteus)* can attack cardiolipin with liberation of 2 mol of fatty acid per mol of phosphorus. This result argues for an *sn*-3 configuration for the native cardiolipin molecule.

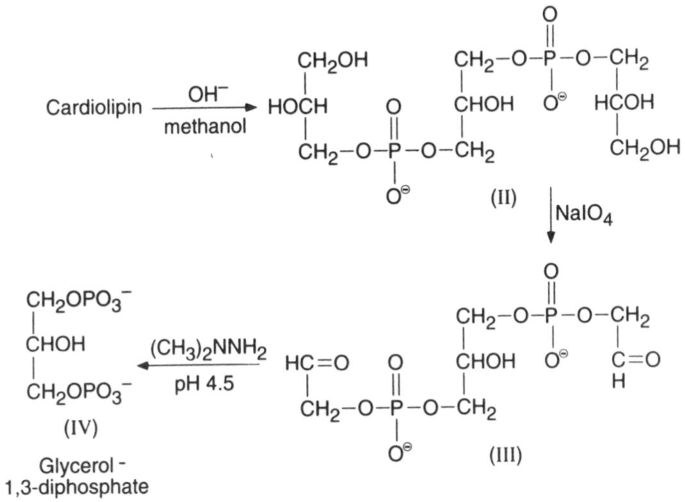

FIGURE 6-7. Periodic acid oxidation of deacylated cardiolipin.

Mass Spectrometric Examination

This powerful technique will provide considerable information on the structure of cardiolipin. If FAB-MS is used in the negative ion mode, cardiolipin shows an M_r-H at 1448, assuming 18:2 acyl residues; monolysocardiolipin with only 18:2 residues will show a mass ion at 1186 and the dilyso form a mass ion at 924 (with only 18:2 acyl residues) (Robinson, 1990).

COMMENTS. The identification of cardiolipin in a biological sample poses a challenge. In particular, the small amounts of this compound normally found in mammalian cells, its sensitivity to air oxidation, and its ability to form salts with various cations (Na^+, Ca^{2+}, and others) are factors impacting on its isolation and characterization. If cardiolipin is detected on thin-layer chromatography by P and TNS spray, this would indicate sufficient sample for a mass spectrometric examination. Whenever possible, it is suggested that this potent methodology be used to establish the structural features of the compound under study. There are many well-qualified, chemically knowledgeable mass spectrometrists available, and establishment of a rapport with them would constitute an invaluable resource. The costs for these analyses are very reasonable considering the important information that is derived from such an examination.

Sphingosine-1-Phosphoric Acid

More than 25 years ago, Stoffel et al. (1970) reported the formation of sphingosine-1-P in erythrocytes. This metabolite of sphingosine can be regarded as *trans*-4-sphingenine-1-P (Figure 6-8).

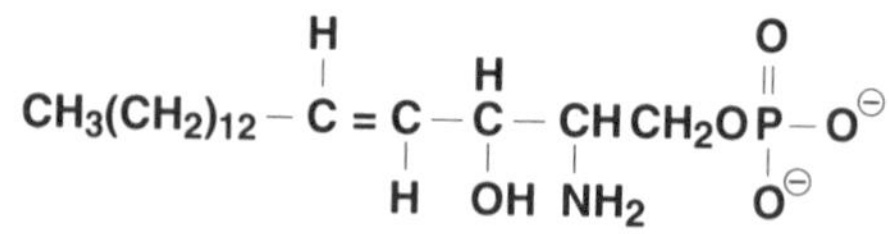

$$CH_3(CH_2)_{12} - \overset{\overset{\displaystyle H}{|}}{C} = \overset{\overset{\displaystyle H}{|}}{\underset{\underset{\displaystyle H}{|}}{C}} - \overset{}{\underset{\underset{\displaystyle OH}{|}}{C}} - \overset{}{\underset{\underset{\displaystyle NH_2}{|}}{CH}} CH_2 O \overset{\overset{\displaystyle O}{\|}}{\underset{\underset{\displaystyle O^\ominus}{|}}{P}} - O^\ominus$$

FIGURE 6-8. Structural formula for sphingosine-1-P.

Renewed interest in this compound has emerged from the observations that sphingosine *in vivo* could be a protein kinase C inhibitor (modulator) and the formation of sphingosine-1-P could be regarded as an attenuation process involved in the control of this inhibition (Van Veldhoven et al., 1989). Later, Zhang et al. (1991) presented compelling data in quiescent 3T3 cells on the formation of sphingosine-1-P and on its involvement in cellular proliferation and in mobilization of intracellular calcium ions. The action of these compounds appears similar to those recorded for the lysophosphatidic acids discussed earlier in this chapter. The interesting possibility arises that a hydrophobic residue (represented by the long-chain hydrocarbon component) linked to a phosphoric acid moiety forms the basic structure required for biological activity.

Isolation

The extraction of this lipid from a cellular preparation can be accomplished through the use of an acidified chloroform–methanol mixture as described earlier for isolation of the lysophosphatidic acids and other acidic phospholipids. The chloroform-rich phase represented the total lipid and can be studied further as described in the following section.

Purification and General Identification

The preferred route to recovery of sphingosine-1-P is through two-dimensional thin-layer chromatography on preparative or analytical plates, depending on the amount of the compound present. Silica gel 60G-coated plates are effective, together with a solvent system of chloroform–methanol–ammonia (28%); (65:37:7, v/v) in the first direction and chloroform–acetone–methanol–acetic acid–water (45:20:10:13:5, v/v) in the second direction. The location of sphingosine-1-P-can be achieved by using a phosphorus spray, the ninhydrin reagent, periodate cleavage (Schiff base reaction), and/or the TNS spray. It is mandatory, however, to run standards on a second plate, since it is possible that a presumed sphingosine-1-P spot may contain other components such as the lysophosphatidic acids.

Further characterization of the sphingosine-1-P, eluted from the two-dimensional plate by acidified chloroform–methanol, is obtained by chromatographing the extracted material in a single dimension on another silica gel plate. The solvent system of choice is that of butanol–water–acetic acid (3:1:1, v/v) as described by Zhang et al. (1991).

Enzymatic Synthesis of Sphingosine-1-P:
An *In Vitro* Approach

In order to conduct research studies on this novel lipid mediator, it is necessary to have available a sufficient quantity of purified, well-defined material. This has not been an easy task, but an approach to preparation of sphingosine-1-P in milligram amounts was outlined by Van Veldhoven et al. (1989). In this procedure, a commercial preparation of sphingosylphosphocholine (which contains a free amine group due to removal of a fatty acyl residue from initial substrate, sphingomyelin) is incubated with phospholipase D in an ammonium acetate buffer at pH 8.0 for 1 hr. An insoluble reaction product is collected and subjected to purification by dissolution in water at room temperature followed by cooling to 4°C. A precipitate forms again and is collected and treated in a similar manner as above except that acetone is used as the solvent.

Characterization of Product

Silica gel 60G-coated plates can be used with thin-layer chromatography, and a suitable solvent system is butanol–acetic–water (6:2:2, v/v). If the product is pure sphingosine-1-P, it should migrate with an R_f value of 0.48. It will give a positive reaction to the phosphorus and ninhydrin reagents, as well as exhibiting a positive reaction to iodine vapors.

NUCLEAR MAGNETIC RESONANCE (NMR) SPECTRUM.   Carbon-13 proton NMR of the starting material used in the preparation of the sphingosine-1-P (i.e., sphingosylphosphocholine) showed the presence of two stereochemical forms, the D-*erythro* and the L-*threo,* in a ratio of 72:28. These isomers were formed during the acid hydrolysis of sphingomyelin (D-*erythro* form only) in which the fatty acid amide bond is cleaved to yield the free-amine-containing sphingosylphosphocholine together with the liberation of a free fatty acid. This type of information should be of decided interest to any investigator studying the importance of a specific stereochemical form in the action of sphingosine-1-P.

MASS SPECTRAL PATTERN.   Only a limited examination of sphingosine-1-P through the use of mass spectrometry has been reported. Using FAB-MS and a thioglycerol matrix, Van Veldhoven et al. (1989) noted the occurrence of a mass ion m/z at 380, which is expected for sphingosine-1-P. Other ions ranging from 264 to 121 were reported, but no structural assignments were made.

Concluding Remarks (The Finale)

The dramatic developments in the past several years in the field of signal transduction processes and other membrane-associated reactions has mirrored the unprecedented awareness of scientists for membrane phospholipids. For many years, studies on the chemical and biochemical behavior of cellular phospholipids were restricted to a relatively few laboratories with specialized interest in these compounds. Part of the latter interest was focused on the isolation, purification, and structure proof of membrane phospholipids. No doubt this was of penetrating interest to these dedicated souls, but the subject matter did not have widespread appeal and did not attract a dedicated following. There was a rather rapid resurgence of interest in these compounds when they were shown to be involved in the signal transduction process.

It is now clear that an understanding of the chemical structure of the major (masswise) lipid in mammalian membranes (i.e., the phospholipids) is of paramount importance in cellular reactions. Such knowledge plays a defining role in interpreting their role in physiological as well as pharmacological events. The fact that cellular response to a variety of agonists intimately involves phospholipids has lured scientists with diverse backgrounds to include these fascinating compounds in their experimental protocols. While this of course brings warmth to the hearts of lipid biochemists, many of these scientists have not had any formal training with these compounds. Particularly true is the fact that graduate study in the biological sciences at most institutions of higher learning does not include any formal contact (as a required course, or special seminars) with lipid biochemistry. As a result, many investigators new to the field of signal transduction may have had a limited exposure to lipid chemistry and biochemistry and may not be aware of the subtleties of this area.

Thus, given this backdrop, the decision was made to prepare a gentle guide to the chemistry of phospholipids. As stated in the preface, this book was not to be construed as a tome (or an encyclopedia of facts), but rather as an introduction to the chemistry of these unique compounds. It is hoped that the reader will utilize the material presented here as a basis for further study, whether in the laboratory or in the library, on this remarkable group of compounds. While the subject matter was restricted to phospholipids present in mammalian cells, there is no reason why the same approaches and logic cannot be applied to similar compounds present in plants, bacteria, and fish. There is much to be learned about the behavior of the phospholipids in cellular metabolism, cell-cell interactions, initiation of signals in a cell, and the overall functioning of the cell. Future developments in these areas undoubtedly will require a significant revision with regard to the role of phospholipids in biological systems.

REFERENCES

Chapter One

Baer, E. and Buchnea, D. (1959) Synthesis of L-α-(dioleoyl)-cephalin, with a comment on the stereochemical designation of glycerol phosphatides, *J. Am. Chem. Soc.* **81,** 1758–1762.

Berridge, M. J. and Irvine, R. F. (1984) Inositol trisphosphate: a novel second messenger in cellular signal transduction, *Nature* **312,** 315–321.

Braquet, P., Touqui, L., Shen, T. Y., and Vorgaftig, B. B. (1987) Perspectives in platelet-activating factor research, *Pharmacol. Rev.* **39,** 97–145.

Cahn, R. S., Ingold, C. K., and Prelog, V. (1966) Specification of molecular chirality, *Angew. Chem.* **5,** 385–415.

Demopoulos, C. A., Pinckard, R. N., and Hanahan, D. J. (1979) Platelet activating factor (PAF): evidence for 1-*O*-alkyl-1-acetyl-*sn*-glyceryl-3-phosphorylcholine as the active component (a new class of lipid chemical mediators), *J. Biol. Chem.* **254,** 9355–9358.

Exton, J. (1990) Signalling through phosphatidyl choline breakdown, *J. Biol. Chem.* **265,** 1–4.

Hanahan, D. J. (1986) Platelet activating factor. A biologically active phosphoglyceride, *Annu. Rev. Biochem.* **55,** 483–510.

Hanahan, D. J., Demopoulos, C. A., Liehr, J., and Pinckard, R. N. (1980) Identification of platelet activating factor isolated from rabbit basophils as actylglyceryletherphosphorylcholine, *J. Biol. Chem.* **255,** 5514–5516.

Hirschmann, H. (1960) The nature of substrate asymmetry in stereoselective reactions, *J. Biol. Chem.* **235,** 2762–2767.

Hokin, M. R., and Hokin, L. L. (1953) Enzyme secretion and the incorporation of ^{32}P into phospholipids of pancreas slices, *J. Biol. Chem.* **203,** 967–977.

IUPAC-IUB Commission on Biochemical Nomenclature (CBN). (1967) The nomenclature of lipids, *Eur. J. Biochem.* **2,** 127–131.

195

Kuksis, A., editor. (1978) Fatty Acids and glycerides, in *Handbook of Lipid Research*, Plenum Press, New York, pp. 1–121.

Markely, K. S., editor. (1960, 1961, 1964) *Fatty Acids*, parts 1, 2, and 3, Interscience Publishers, New York.

Michell, R. H. (1975) Inositol phospholipids and cell surface receptor function, *Biochim. Biophys. Acta* **415**, 81–147.

Nelson, D. R. and Hanahan, D. J. (1985) Phospholipid and detergent effects on (Ca + Mg) ATPase purified from human erythrocytes, *Arch. Biochem. Biophys.* **236**, 720–730.

Nishizuka, Y. (1992) Intracellular signaling by hydrolysis of phospholipids and activation of protein kinase C, *Science* **258**, 607–614.

Ogston, A. G. (1948) Interpretation of experiments on metabolic processes, using isotopic tracer elements, *Nature* **162**, 963.

Ojima-Uchiyama, A., Masuzawa, Y., Sugiura, T., Waku, K., Saito, H., Yui, Y., and Tamioka, H. (1988) Phospholipid analyses of human eosinophils: high levels of alkylacyglycerophosphocholine (PAF precursor) *Lipids* **23**, 815–817.

Pinckard, R. N., Farr, R. S., and Hanahan, D. J. (1979) Physicochemical and functional identity of rabbit platelet-activating factor (PAF) release *in vivo* during IgE anaphylaxis with PAF Released *in vitro* from IgE sensitized basophils, *J. Immunol.* **123**, 1847–1857.

Snyder, F. (1982) Platelet activating factor (PAF): a novel type of phospholipid with diverse biological properties. *Annu. Rep. Med. Chem.* **17**, 243–252.

Ways, P., Reed, C. F., and Hanahan, D. J. (1963) Red cell and plasma lipids in acanthocytosis, *J. Clin. Invest.* **42**, 1248–1260.

Chapter Two

Bligh, E. G. and Dyer, W. (1959) A rapid method of total lipid extraction and purification, *Can. J. Biochem. Physiol.* **37**, 911–917.

Bloor, W. R. (1928) The determination of small amounts of lipid in blood, *J. Biol. Chem.* **77**, 53–73.

Carter, T. P. and Kanfer, J. N. (1973) Methodology for separation of gangliosides from potential water-soluble precursors, *Lipids* **8**, 537–548.

Danielli, J. F. (1982) Experiment, hypothesis and theory in the development of concepts of cell membrane structure, 1930–1970, in *Membranes and Transport*, Vol. 1, A. N. Martonosi, editor, Academic Press, New York, pp. 1–14.

Folch, J., Lees, M., and Sloane-Stanley, G. H. (1951) A simple method for the isolation and purification of total lipids from animal tissues, *J. Biol. Chem.* **226**, 497–509.

Hakomori, S.-i. (1983) Sphingolipid biochemistry, in *Handbook of Lipid Research*, J. N. Kanfer, and S.-i. Hakomori, editors, Plenum Press, New York, pp. 11–37.

Halliwell, B. and Chirico, S. (1993) Lipid peroxidation: its mechanism, measurement and significance, *Am. Soc. Clin. Nutrition* **57**(suppl.), 7155–7255.

Hanahan, D. J. and Chaikoff, I. L. (1947) A new phospholipide-splitting enzyme specific for the ester linkage between the nitrogenous base and the phosphoric acid group, *J. Biol. Chem.* **169**, 699–705.

Petty, H. R. (1993) *Molecular Biology of Membranes: Structure and Function*, Plenum Press, New York.

Stimmel, J. B., Deschenes, R. J., Volker, C., Stock, J., and Clarke, S. (1990) Evidence for an S-farnesyl cysteine methyl ester at the carboxyl terminus of the *Saccharomyces* RAS 2 protein, *Biochemistry* **29**, 9651–9659.

Towler, D. A., Gordon, S. I., Adams, S. P., and Glaser, L. (1988) The biology and enzymology of eukaryotic protein acylation, *Annu. Rev. Biochem.* **57**, 69–99.

Wells, M. A. and Dittmer, J. C. (1963) The use of Sephadex for the removal of nonlipid contaminants from lipid extracts, *Biochemistry* **2**, 1259–1263.

Wirtz, K. W. A., Packer, L., Gustafsson, A. J., Evangelopoulos, A. E., and Changeux, J. P. editors, *New Developments in Lipid-Protein Interactions and Receptors Function* (1993) Plenum Press, New York.

Chapter Three

Bartlett, G. R. (1959) Phosphorus assay in column chromatography, *J. Biol. Chem.* **234**, 466–488.

Berger, H., Jones, P., and Hanahan, D. J. (1972) Structural studies on lipids of tetrahymena pyriformis, *Biochim. Biophys. Acta* **260**, 617–629.

Billah, M. M. and Lapetina, E. G. (1982) Degradation of phosphatidylinositol-4,5 bisphosphate is insensitive to Ca^{2+} mobilization is stimulated platelets, *Biochem. Biophys. Res. Commun.* **109**, 217–222.

Chen, S. S.-H. and Kou, A. Y. (1952) Improved procedure for the separation of phospholipids by high performance liquid chromatography, *J. Chromatogr.* **227**, 25–31.

Dittmer, J. C. and Lester, R. L. (1964) A simple specific spray for the detection of phospholipids on thin layer chromatography, *J. Lipid Res.* **5**, 126–127.

Gonzalez-Sastro, F. and Folch-Pi, J. (1968) Thin layer chromatography of the phosphoinositides, *J. Lipid Res.* **9**, 532–533.

Hakomori, S.-i. (1983) Chemistry of Glycosphingolipids in *Sphingolipid Biochemistry,* J. N. Kanfer and S.-i. Hakomori, editors, Plenum Press, New York, p. 35.

Hanahan, D. J., Turner, M. B., and Jayko, M. E. (1951) The isolation of egg phosphatidylcholine by an adsorption column technique, *J. Biol. Chem.* **192**, 623–628.

Hanahan, D. J. and Weintraub, S. T. (1985) Platelet activating factor. Isolation, identification and assay, in *Methods and Biochemical Analysis,* D. Glick, editor, John Wiley & Sons, New York, pp. 195–219.

Jones, M., Keenan, R. W., and Horowitz, P. (1982) Use of 6-*p*-toluidine-2-napthalene sulfonic acid to quantitate lipids after thin layer chromatography, *J. Chromatogr.* **235**, 522–524.

Plattner, R. D. (1981) High performance liquid chromatography of triglycerides, in *Methods of Enzymology,* Vol. 72, S. P. Colowick and N. O. K. Kaplan, editors, Academic Press, New York, pp. 21–34.

Porter, N. A. and Weenan, H. (1981) High performance liquid chromatographic separation of phospholipids and phospholipid oxidation products, in *Methods of Enzymology,* Vol. 72, S. P. Colowick and N. O. K. Kaplan, editors, Academic Press, New York, pp. 34–40.

Shukla, S. D. and Hanahan, D. J. (1982) AGEPC (platelet activating factor) induced stimulation of rabbit platelets: effects on phosphatidyl inositol, di- and tri- phosphoinositides and phosphatidic acid metabolism, *Biochem. Biophys. Res. Commun.* **106**, 697–703.

Tokumura, A., Kramp, W., and Hanahan, D. J. (1986) Alkylacetylglycerophospho-choline effects on the metabolism of phospholipids in rabbit platelets: effects of extracellular Ca^{2+} and prostacyclin, *Arch. Biochem. Biophys.* **247**, 403–413.

Touchstone, J. C. (1992) *Practice of Thin Layer Chromatography,* John Wiley & Sons, New York.

Watson, J. T. (1985) *Introduction to Mass Spectrometry,* 2nd ed., Raven Press, New York, pp. 207–221.

Wells, M. A. and Dittmer, J. C. (1963) The use of Sephadex for the removal of contaminants from lipid extracts, *Biochemistry* **2**, 1259–1265.

Chapter Four

Albro, P. W. and Dittmer, J. C. (1968) Determination of the distribution of the aliphatic groups of glyceryl ethers by gas-liquid chromatography of the diacetyl and derivative, *J. Chromatogr.* **38**, 230–239.

Allgyer, T. J. and Wells, M. A. (1979) Phospholipase D from Savoy cabbage: purifica-tion and preliminary kinetic characterization, *Biochemistry* **18**, 5348–5353.

Arveldano, M. I., van Rollins, M., and Horrocks, L. A. (1983) Separation and quantitation of free fatty acid and fatty acid methyl esters by reverse phase high pressure liquid chromatography, *J. Lipid Res.* **24**, 83–93.

Barak, A. J. and Toma, D. J. (1981) Determination of choline, phosphorylcholine and betaine methods in enzymology, *Lipids* **72** (part D), 287–292.

Brockerhoff, H. and Yurbowski, M. (1965) Simplified preparation of L-α-glyceryl phosphorylcholine, *Can. J. Biochem.* **43**, 1777.

Bruzik, K. S. and Tsai, M.-D. (1991) Phospholipase stereospecificity at phosphorus, in *Methods in Enzymology,* Vol. 197, E. A. Dennis, editor, Academic Press, New York, pp. 258–269.

Bruzik, K., Jiang, R. T., and Tsai, M.-D. (1983) Phospholipids chiral at phosphorus. Preparation and spectral properties of chiral thiophospholipids, *Biochemistry* **22**, 2478–2486.

Bugaut, M., Kuksis, A., and Myher, J. J. (1985) Loss of stereospecificity of phospho-lipase C and D upon introduction of a 2-alkyl group into *rac*-1,2 diacylglycero-3-phosphocholine, *Biochim. Biophys. Acta* **835**, 304–314.

Carter, H. E. and Gaver, R. C. (1967) Improved reagent for trimethylsilyation of sphingolipid bases, *J. Lipid Res.* **8**, 391–395.

Chacko, G. K. and Hanahan, D. J. (1968) Chemical synthesis of 1-*O*-(D)- and 3-O-(L)-glyceryl monoethers, diethers and derivatives: glyceride, monoester phospholipids and diether phospholipids, *Biochim. Biophys. Acta* **164**, 252–271.

Cymerman-Craig, J. and Hamon, D. P. G. (1965) Studies directed toward the synthe-sis of plasmalogens II (±) *cis*- and *trans*-3/*n*-hexadec-1′-enyloxyl-1,2-propanediol, *J. Org. Chem.* **30**, 4168–4175.

Cymerman-Craig, J., Hamon, D. P. G., Purushothaman, K. K., Roy, S. K., and Lands, W. E. M. (1966), Optical rotatory dispersion and absolute configuration. V. Absolute configuration of natural plasmalogens, *Tetrahedron* **23**, 175–178.

Debuch, H. and Seng, P. (1972) The history of ether-linked lipids through 1960, in *Ether Lipids: Chemistry and Biology,* F. Snyder, editor, Academic Press, New York, pp. 1–24.

de Haas, G. H. and van Deenen, L. L. M. (1961) Synthesis of enantioimeric mixed-acid phosphatides, *Rec. Trav. Chim.* **80**, 951–970.

Dennis, E. A. (1991) Phospholipases, in *Methods in Enzymology*, Vol. 197, E. A. Dennis, editor, Academic Press, New York, pp. 3–640.

Dixon, J. S. and Lipkin, D. (1954) Spectrophotometric determination of vicinal glycols. Application to the determination of ribofuranosides, *Anal. Chem.* **26,** 1092–1093.

Egge, H. (1983) Mass spectrometry of ether lipids, in *Ether Lipids. Biochemical and Biophysical Aspects*, H. K. Mangold and F. Paltauf, editors, Academic Press, New York, pp. 17–47.

Eibl, H. (1980) Synthesis of glycerophospholipids, *Chem. Phys. Lipids* **26,** 405–429.

Eibl, H. and Kovatchev, S. (1981) Preparation of phospholipids and their analogs by phospholipase D, in *Methods in Enzymology*, Vol. 72, J. M. Lowenstein, editor, Academic Press, New York, pp. 632–639.

El-Sayed, M. Y., DeBose, C. D., Coury, L. A., and Roberts, M. F. (1985) Sensitivity of phospholipase C *(Bacillus cereus)* activity to phosphatidylcholine structural modifications, *Biochim. Biophys. Acta* **837,** 325–335.

Folch-Pi, J., Lees, M., and Sloane-Stanley, G. H. (1957) A simple method for the isolation and purification of total lipids from animal tissues, *J. Biol. Chem.* **226,** 497–509.

Gassama-Diagne, A., Fauvel, J., and Chap, H. (1989) Purification of a new, calcium-independent, high molecular weight phospholipase A$_2$/lysophospholipase (phospholipase B) from guinea pig intestinal brush-border membrane, *J. Biol. Chem.* **264,** 9470–9475.

Goldwhite, H. (1981) *Introduction to Phosphorus Chemistry*, Cambridge University Press, Cambridge, England.

Gottfried, E. L. and Rapport, M. M. (1962) The biochemistry of plasmalogens I: Isolation and characterization of phosphatidal choline, a pure native plasmalogen, *J. Biol. Chem.* **237,** 329–333.

Gupta, C. M., Radhakrishnan, R., and Khorana, H. G. (1977) Glycerophospholipid synthesis: improved general method and new analogs containing photoactivable groups, *Proc. Natl. Acad. Sci. USA* **74,** 4315–4319.

Hakomori, S.-i. (1983) Chemistry of glycosphingolipids, in *Handbood of Lipid Research*, Vol. 3, J. N. Kanfer and S. I. Hakomori, Plenum Press, New York, pp. 1–165.

Hanahan, D. J. (1952) The enzymatic degradation of phosphatidyl choline in diethyl ether, *J. Biol. Chem.* **195,** 199–206.

Hanahan, D. J. (1961) Sphingomyelin, *Biochem. Prep.* **8,** 121–124.

Hanahan, D. J. and Chaikoff, I. L. (1947) A new phospholipide-splitting enzyme specific for the ester linkage between the nitrogenous base and the phosphoric acid, *J. Biol. Chem.* **169,** 699–705.

Hanahan, D. J. and Weintraub, S. T. (1985) Platelet-activating factor. Isolation, identification and assay, in *Methods of Biochemical Analysis*, Vol. 31, D. Glick, editor, John Wiley & Sons, New York, pp. 195–219.

Hanahan, D. J., Ekholm and Jackson, C. M. (1963) Studies on the structure of the glyceryl ether and the glycerylether phospholipids of bovine erythrocytes, *Biochemistry* **2,** 630–641.

Hanahan, D. J., Nouchi, T., Weintraub, S. T., and Olson, M. S. (1990) Novel route to preparation of high purity lysoplasmenylethanolamine, *J. Lipid Res.* **31,** 2113–2117.

Holub, B. J. and Celi, B. (1984) Evaluation of the fatty acid selectivity of a phosphatidyl inositol-specific cytosolic phospholipase C from pig and human platelets, *Can. J. Biochem. Cell. Biol.* **62**, 115–120.

Huang, K.-S., Li, S., and Low, M. G. (1991) Glycosylphosphatidyl-specific phospholipase D, in *Methods in Enzymology*, E. A. Dennis, editor, Academic Press, New York, pp. 567–579.

Jiang, R.-T., Shyy, Y.-J., and Tsai, M.-D. (1984) Phospholipids chiral at phosphorus. Absolute configuration of chiral thiophospholipids and stereospecificity of phospholipase D, *Biochemistry* **23**, 1661–1667.

Kanfer, J. N. (1983) Sphingolipid metabolism, in *Handbook of Lipid Research*, Vol. 3, J.N. Kanfer and S.-i. Hakomori, editors, Plenum Press, New York, pp. 167–247.

Kanfer, J. N. and Hakomori, S.-i. (1983) Sphingolipid biochemistry, in *Handbook of Lipid Research*, Vol. 3, J. N. Kanfer and S.-i. Hakomori, editors, Plenum Press, New York, pp. 1–485.

Karlsson, K.-A. (1970) Sphingolipid long chain bases, *Lipids* **5**, 878–889.

Karlsson, K.-A. and Pascher, I. (1971) Thin-layer chromatography of ceramide, *J. Lipid Res.* **12**, 466–472.

Karnovsky, M. L. and Brumm, A. F. (1955) Studies of naturally occurring α-glycerol ethers, *J. Biol. Chem.* **216**, 689–701.

Kennerly, D. A. (1991) Quantitative analysis of water-soluble products of cell associated phospholipase C and phospholipase D-catalyzed hydrolysis of phosphatidyl choline, in *Methods in Enzymology*, Vol. 197, E. A. Dennis, editor, Academic Press, New York, pp. 191–197.

Kobayashi, M. and Kanfer, J. N. (1991) Solubilization and purification of rat tissue phospholipase D, in *Methods in Enzymology*, Vol. 197, E. A. Dennis, editor, Academic Press, New York, pp. 575–583.

Kuksis, A. (1978) Separation and determination of structure of fatty acids, in *Handbook of Lipid Research*, Vol. 1, A. Kuksis, editor, Plenum Press, New York, pp. 1–121.

Kumar, R., Weintraub, S. T., McManus, L. M., Pinckard, R. N., and Hanahan, D. J. (1984) A facile route to semi-synthesis of acetyl glycerylether phorylethanolamine and its choline analogue, *J. Lipid Res.* **25**, 198–208.

Mangold, H. K. and Paltauf, F., editors. (1983) *Ether Lipids. Biochemical and Biomedical Aspects*, Academic Press, New York, pp. 1–439.

Mangold, H. K. and Totani, N. (1983) Procedures for the analysis of ether lipids, in *Ether Lipids. Biochemical and Biomedical Aspects*, H. K. Mangold, and F. Paltauf, editors, Academic Press, New York, pp. 377–387.

Massing, U. and Eibl, H. (1994) Substrates for phospholipase C and sphingomyelinase from *Bacillus cereus*, in *Lipases*, P. Woolley and S. S. Peterson, editors, Cambridge University Press, Cambridge, England, pp. 225–242.

Matsuzawa, Y. and Hostetler, K. Y. (1980) Properties of phospholipase C isolated from rat liver lysosomes, *J. Biol. Chem.* **255**, 646–652.

Misiorowski, R. I. and Wells, M. A. (1974) The activity of phospholipase A$_2$ in reversed micelle of phosphatidyl-choline in diethyl ether: effect of water and cations, *Biochemistry* **13**, 4921–4927.

Murphy, E. J., Stephens, R., Jurkowitz-Alexander, M., and Horrocks, L. A. (1993) Acidic hydrolysis of plasmalogens followed by high-performance liquid chromatography, *Lipids* **28**, 565–568.

Murphy, R. C. (1993) Mass spectrometry of lipids, in *Handbook of Lipid Research*, Vol. 7, Plenum Press, New York, pp. 1–308.

Murray, R. K. and Narasimhan, R. (1990) Glycoglycerolipids of animal tissues, in *Handbook of Lipid Research*, Vol. 6, M. Kates, editor, Plenum Press, New York, pp. 321–361.

Nishizuka, Y. (1992) Intracellular signaling by hydrolysis of phospholipids and activation of protein kinase C, *Science* **258**, 607–614.

Norton, W. T., Gottfried, E. L., and Rapport, M. M. (1962) The structure of plasmalogens: VII. Configuration of the double bond in the α,β-unsaturated linkage of phosphatidal choline, *J. Lipid Res.* **3**, 456–459.

Paltauf, F. (1983) Chemical synthesis of ether lipids, in *Ether Lipids. Biochemical and Biomedical Aspects*, H. K. Mangold and F. Paltauf, editors, Academic Press, New York, pp. 49–84.

Patel, S. (1967) *The Chemistry of the Ether Linkage*, Interscience, New York.

Poon, P. and Wells, M. A. (1974) Physical studies of egg phosphatidylcholine in diethyl water solutions, *Biochemistry* **13**, 4928–4936.

Quin, L. D. and Verkade, J. G. (1981) *Phosphorus Chemistry*, ACS Symposium Series, American Chemical Society, Washington, D.C.

Reynolds, L. J., Washburn, W. N., Deems, R. A., and Dennis, E. A. (1991) Assay strategies and methods for phospholipases, in *Methods in Enzymology*, Vol. 197, E. A. Dennis, editor, Academic Press, New York, pp. 1–23.

Rhee, S. G., Ryu, S. H., Lee, K. Y., and Cho, K. S. (1991) Assays of phosphoinositide-specific phospholipase C and purification of isozymes from bovine brain, in *Methods in Enzymology*, Vol. 197, E. Dennis, editor, Academic Press, New York, pp. 502–511.

Saito, K. and Hanahan, D. J. (1962) A study of the purification and properties of the phospholipase A_2 of *Crotalus adamanteus* venom, *Biochemistry* **1**, 521–532.

Saito, K., Sugatani, J., and Okumura, T. (1991) Phospholipase B from *Penicillium notatum*, in *Methods in Enzymology*, Vol. 197, E. A. Dennis, editor, Academic Press, New York, pp. 446–456.

Satouchi, K. and Saito, K. (1979) Use of *t*-butyldimethylchlorosilane/imidazole reagent for identification of molecular species of phospholipids by gas–liquid chromatography–mass spectrometry, *Biomed. Mass Spectrom.* **6**, 396–402.

Satouchi, K., Pinckard, R. N., McManus, L. M., and Hanahan, D. J. (1981) Modification of the polar head group of acetyl glyceryl ether phosphorylcholine and subsequent effects upon platelet activation, *J. Biol. Chem.* **256**, 4425–4432.

Scott, D. L., White, S. P. Otwinowski, Z., Yuan, W., Gelb, M. H., and Sigler, P. B. (1990) Interfacial catalysis: the mechanism of phospholipase A2, *Science* **250**, 1541–1546.

Siggia, S. and Edsberg, R. L. (1948) Iodometric determination of vinyl alkyl ether, *Anal. Chem.* **20**, 762–763.

Sloane-Stanley, G. (1953) Anaerobic reactions of phospholipins in brain suspensions, *Biochem. J.* **53**, 613–619.

Sloane-Stanley, G. H. (1967) Simple procedure for the estimation of very small amounts of nitrogen in lipids, *Biochem. J.* **104**, 293–295.

Snyder, F. (1972) *Ether Lipids, Chemistry and Biology*, Academic Press, New York.

Snyder, F. and Stephens, N. (1959) Siimplified spectrophotometric determination of ester groups in lipids, *Biochim. Biophys. Acta* **34**, 244–245.

Stavinoha, W. B. and Weintraub, S. T. (1974) Estimation of choline and acetylcholine in tissue by pyrolysis gas chromatography, *Anal. Chem.* **46**, 757–760.

Sugiura, T., Nakajima, M., Sekiguchi, N., Nakagawa, Y., and Waku, K. (1983)

Different fatty chain components of alkenylacyl, alkylacyl and di-acylphospholipids in rabbit aveolar macrophages: high amounts of arachidonic acid in ether phospholipids, *Lipids* **18**, 125–129.

Suzuki, K. (1965) The pattern of mammalian brain gangliosides—II. Evaluation of the extraction procedures, post-mortem changes and the effect of formalin preservation, *J. Neurochem.* **12**, 629–638.

Thompson, G. A. and Kapoulas, V. M. (1968) Preparation and assay of glyceryl *ethers,* in *Methods in Enzymology,* Vol. 14, J. Lowenstein, editor, Academic Press, New York, pp. 668–678.

Waite, M. (1987) The Phospholipases, in *Handbook of Lipid Research,* Vol. 5, Plenum Press, New York, pp. 1–332.

Waku, K. and Nakagawa, Y. (1972) Hydrolyses of 1-*O*-alkyl, 1-*O*-alkenyl and 1-acyl-2-(1-^{14}C)linoleoyl-glycero-3-phosphoryl choline by various phospholipases, *J. Biochem.* **72**, 149–155.

Warner, H. R. and Lands, W. E. M. (1963) The configuration of the double bond in naturally-occurring alkenyl ether, *J. Am. Chem. Soc.* **85**, 60–64.

Weintraub, S. J., Tokumura, A., and Hanahan, D. J. (1991) Analysis of lysoplasmenylethanolamine by FAB mass spectrometry, in *Proceedings of the 39th ASMS Conference on Mass Spectrometry and Allied Topics,* Nashville, pp. 1033–1034.

Wells, M. A. (1971) Evidence for *O*-acyl cleavage during hydrolysis of 1,2 diacyl-*sn*-glycero 3-phosphocholine by the phospholipase A$_2$ of *Crotalus adamantus, Biochim. Biophys. Acta* **248**, 80–86.

Wells, M. A. and Hanahan, D. J. (1969) Studies on phospholipase A. I. Isolation and characterization of two enzymes of *Crotalus adamanteus* venom, *Biochemistry* **8**, 414–424.

Wilson, E., Wang, E., Mullins, R. E., Uhlinger, D. J., Liotta, D. C., Lambeth, J. D., and Merrill, A. H. (1988) Modulation of the free sphingosine levels in human neutrophils by phorbol esters and other factors, *J. Biol. Chem.* **263**, 9304–9309.

Wittenberg, J., Korey, S. R., and Swenson, F. H. (1956) The determination of higher fatty aldehydes in tissues, *J. Biol. Chem.* **219**, 39–47.

Yang, S. F., Freer, S., and Benson, A. A. (1967) Transphosphatidylation by phospholipase D, *J. Biol. Chem.* **242**, 477–484.

Chapter Five

Adams, R. F. (1974) Determination of amino acid profile in biological samples by gas chromatography, *J. Chromatogr.* **95**, 189–212.

Ballou, C. E. and Pizer, L. I. (1960) The absolute configuration of the *myo*-inositol-1-phosphates and a confirmation of the bornesitol configuration, *J. Am. Chem. Soc.* **82**, 3333–3335.

Brockerhoff, H. and Hanahan, D. J. (1959) Studies on naturally occurring phosphoinositides, *J. Am. Chem. Soc.* **81**, 2591–2592.

Brown, D. M. and Stewart, J. C. (1966) The structure of triphosphoinositide from beef brain, *Biochim. Biophys. Acta* **125**, 413–421.

Chen, S. S-H., Kou, A. Y., and Chen, H-H. Y. (1983) Quantitative analysis of aminophospholipids by high-performance liquid chromatography using succinimidyl 2-naphthoxyacetate as a fluorescent label, *J. Chromatogr.* **276**, 37–44.

Dawson, R. M. C. and Dittmer, J. C. (1961) Evidence for the structure of brain triphosphorinositide from hydrolytic degradation studies, *Biochem. J.* **81**, 540–545.

Dawson, R. M. C. and Freinkel, N. (1961) The distribution of free-*meso*-inositol in mammalian tissues, including some observations on the lactating rat, *Biochem. J.* **78**, 606–610.

Dawson, R. M. C., Freinkel, N., Jungawala, F. B., and Clarke, N. (1971) The enzymic formation of myoinositol 1:2 cyclic phosphate from phosphatidyl inositol, *Biochem. J.* **127**, 605–607.

Dean, N. M. and Moyer, J. D. (1987) Separation of multiple isomers of inositol phosphate formed in GH$_3$ cells, *Biochem. J.* **242**, 361–366.

Devane, W. A., Hanus, L., Breuer, A., Pertwee, R. G., Stevenson, L. A., Griffin, G., Gibson, D., Mandelbaum, A., Etinger, A., and Mechoulam, R. (1992) Isolation and structure of a brain constituent that binds to the cannabinoid receptor, *Science* **258**, 1946–1949.

Dittmer, J. C. and Dawson, R. M. C. (1961) The isolation of a new lipid, triphosphoinositide, and monophosphoinositide from ox brain, *Biochem. J.* **81**, 535–540.

Dugan, L. L., Demedink, P., Pendley, C. E., and Horrocks, L. A. (1986) Separation of phospholipids by high-performance liquid chromatography: all major classes, including ethanolamine and choline plasmalogens, and most minor classes, including lysophosphatidyl ethanolamine, *J. Chromatogr.* **378**, 317–327.

Folch, J. (1942) Brain cephalin, a mixture of phosphatides. Separation from it of phosphatidyl serine, phosphatidyl ethanolamine and a fraction containing an inositol phosphatide, *J. Biol. Chem.* **146**, 35–44.

Folch, J. (1949) Brain diphosphoinositide, a new phosphatide having inositol metadiphosphate as a constituent, *J. Biol. Chem.* **177**, 505–519.

Garvin, J. E. and Karnovsky, M. L. (1956) The titration of some phosphatides and related compounds in a non-aqueous medium, *J. Biol. Chem.* **221**, 211–222.

Grado, C. and Ballou, C. E. (1960) Myo-inositol phosphates from beef brain phosphoinositide, *J. Biol. Chem.* **235**, PC23–24.

Grado, C. and Ballou, C. E. (1961) Myo-inositol phosphate obtained by alkaline hydrolysis of beef brain phosphoinositide, *J. Biol. Chem.* **236**, 54–60.

Hara, A. and Radin, N. S. (1978) Lipid extraction of tissues with a low toxicity solvent, *Anal. Biochem.* **90**, 420–426.

Hokin, M. R. and Hokin, L. E. (1953) Enzyme secretion and the incorporation of P^{32} into phospholipids, *J. Biol. Chem.* **203**, 967–977.

Hunkapillar, M. W., Strickler, J. E., and Wilson, K. J. (1984) Contemporary methodology for protein structure determination, *Science* **226**, 304–311.

Irvine, R. F., Anggard, E. E., Letcher, A. J., and Downes, C. P. (1985) Metabolism of inositol 1,4,5 tris phosphate in rat parotid glands, *Biochem. J.* **229**, 505–511.

Juneja, L. R., Kazuoka, T., Goto, N. Yamane, T., and Shimizu, S. (1989) Conversion of phosphatidylcholine to phosphatidylserine by various phospholipases D in the presence of L- and D-serine, *Biochim. Biophys. Acta* **1003**, 277–283.

Kataoka, H. Sakiyama, N. Maeda, M., and Makita, M. (1989) Determination of phosphoethanolamine in animal tissues by gas chromatography with flame photometric detection, *J. Chromatogr.* **494**, 283–288.

Kates, M. (1972) *Techniques of Lipidology: Isolation, Analysis and Identification of Lipids,* North-Holland, Amsterdam.

Kerwin, J. L., Tuininga, A. R., and Ericsson, L. H. (1994) Identification of molecular species of glycerophospholipids and sphingomyelin using electrospray mass spectrometry, *J. Lipid Res.* **35**, 1102–1114.

Majerus, P. W., Connolly, T. M., Deckym, H., Ross, J. S., Brass, T. E., Ishii, H., Bansal, V. S., and Wilson, D. B. (1986) The metabolism of phosphoinositide-derived messenger molecules, *Science* **234**, 1519–1525.

McMaster, C. R. and Choy, P. C. (1992) The determination of tissue ethanolamine levels by reverse-phase high performance liquid chromatography, *Lipids* **27**, 560–563.

Michell, R. H., Kirk, C. J., Jones, L. M., Downes, C. P., and Creba, J. A. (1981) The stimulation of inositol lipid metabolism that accompanies calcium mobilization in stimulated cells: defined characteristics and unanswered questions, *Philos. Trans. R. Soc. (London B)* **296**, 123–127.

Michell, R. H. (1992) Inositol lipids in cellular signalling mechanisms, *TIBS* **17**, 274–276.

Nishizuka, Y. (1992) Intracellular signaling by hydrolysis of phospholipids and activation of protein kinase C, *Science* **258**, 607–614.

Parthasarathy, R. and Eisenberg, F. (1986) The inositol phospholipids: a stereochemical view of biological activity, *Biochem. J.* **235**, 1–10.

Pizer, F. L. and Ballou, C. E. (1959) Studies on myo-inositol phosphates of natural origin, *J. Am. Chem. Soc.* **81**, 915–921.

Reitz, A. B. (1991) *Inositol Phosphates and Derivatives, Synthesis, Biochemistry and Therapeutic Potential,* ACS Symposia Series, Washington, D.C.

Roberts, W. L., Kim, B. H., and Rosenberg, T. L. (1987) Differences in the glycolipid membrane anchors of bovine and human erythrocyte acetylcholinesterases, *Proc. Natl. Acad. Sci. USA* **84**, 7817–7821.

Schmid, H. H. O., Schmid, P. C., and Natarajan, V. (1990) *N*-Acylated glycerophospholipids and their derivatives, *Prog. Lipid Res.* **29**, 1–43.

Stephens, L. R., Hughes, K. T., and Irvine, R. F. (1991) Pathway of phosphatidylinositol (3,4,5)-trisphosphate synthesis in activated neutrophils, *Nature* **351**, 33–39.

Stephens, L. R., Jackson, T. R., and Hawkins, P. T. (1993) Agonist-stimulated synthesis of phosphatidyl inositol (3,4,5)-trisphosphate: a new intracellular signalling system? *Biochim. Biophys. Acta* **1179**, 27–75.

Streb, H., Irvine, R. F., Berridge, M. J., and Schulz, I.(1983) Release of Ca^{2+} from a nonmitochrondial intracellular store in pancreatic acinar cells by inositol-1,4,5-trisphosphate, *Nature* **306**, 67–69.

Sundler, R. and Akesson, B. (1975) Regulation of phospholipid biosynthesis in isolated rat hepatocytes, *J. Biol. Chem.* **250**, 3359–3367.

Takai, Y., Kishimoto, A., Kikkawa, U., Mori, T., and Nishizuka, Y. (1979) Unsaturated diacylglycerol as a possible messenger for the activation of calcium-activated phospholipid-dependent protein kinase system, *Biochem. Biophys. Res. Commun.* **91**, 1218–1224.

Trayner-Kaplan, A. E., Harris, A. L., Thompson, B., Taylor, P., and Sklar, L. A. (1988) An inositol tetrakis phosphate-containing phospholipid in activated neutrophils, *Nature* **334**, 353–356.

Chapter Six

Balazy, M., Braquet, P., and Bazan, N. G. (1991) Determination of platelet-activating factor and alkyl-ether phospholipids by gas chromatography–mass spectrometry via direct derivatization, *Anal. Biochem.* **196**, 1–10.

Bernhard, W., Linck, M., Creutzberg, H., Postle, A. D., Arning, A., Martin-Carrera,

L., and Sewing, K.-Fr. (1994) High performance liquid chromatographic analysis of phospholipids from different sources with combined fluoresence and ultraviolet detection, *Anal. Biochem.* **220**, 172–180.

Böttcher, C. S. F., van Gent, C. M., and Prus, C. (1961) A rapid and sensitive submicro phosphorus determination, *Anal. Chim. Acta* **24**, 203–204.

Clay, K. L., Sterre, D. O., and Murphy, R. C. (1984) Quantitative analysis of platelet activating factor (AGEPC) by fast atom bombardment mass spectrometry, *Biomed. Mass Spectrum* **11**, 47–49.

Cohen, P. and Derksen, A. (1969) Comparison of phosphalipid and fatty acid composition of human erythrocytes and platelets, *Br. J. Haematol.* **17**, 359–371.

Exton, J. H. (1990) Signaling through phosphatidyl choline breakdown, *J. Biol. Chem.* **265**, 1–4.

Gray, G. M. (1964) The isolation of phosphatidyl glycerol from rat liver mitochondria, *Biochim. Biophys. Acta* **84**, 35–40.

Hanahan, D. J. (1986) Platelet activating factor: a biologically active phosphoglyceride, *Annu. Rev. Biochem.* **55**, 483–509.

Hanahan, D. J. and Weintraub, S. T. (1985) Platelet-activating factor: isolation, identification and assay, in *Methods of Biochemical Analysis*, Vol. 31, D. Glick, editor, John Wiley & Sons, New York, pp. 195–219.

Hanahan, D. J. and Kumar, R. (1987) Platelet activating factor: chemical and biochemical characteristics, *Prog. Lipid Res.* **26**, 1–28.

Haverkate, F. and van Deenen, L. L. M. (1964) The stereochemical configuration of phosphatidyl glycerol, *Biochim. Biophys. Acta* **84**, 106–108.

Haverkate, F. and van Deenen, L. L. M. (1965) Isolation and chemical characterization of phosphatidyl glycerol from spinach leaves, *Biochim. Biophys. Acta* **106**, 78–92.

Heinze, F. J., Linscheid, M., and Heinz, E. (1984) Release of diacylglycerol moieties from various glycosyl diglycerides, *Anal. Biochem.* **139**, 126–133.

Hoch, F. (1992) Cardiolipins and biomembrane function, *Biochim. Biophys. Acta* **1113**, 71–133.

Hovius, R., Lambrechts, H., Nicolay, K., and deKruijff, B. (1990) Improved methods to isolate and subfractionate rat liver mitochondria. Lipid composition of the inner and outer membrane, *Biochim. Biophys. Acta* **1021**, 217–226.

Joo, C. N. and Kates, M. (1969) Synthesis of the naturally occurring phytanyl diether analogs of phosphatidylglycerophosphate and phosphatidylglycerol, *Biochim. Biophys. Acta* **176**, 278–297.

Mauco, G., Chap, H., Simon, M.-F., and Douste-Blazy, X. Y. (1978) Phosphatidic and lyso-phosphatidic acid production in phospholipase C- and thromlin-treated platelets. Possible involvement of a platelet lipase, *Biochimie* **60**, 653–661.

Miwa, M., Hill, C., Kumar, R., Sugatani, J., Olson, M. S., and Hanahan, D. J. (1987) Occurrence of an endogenous inhibitor of platelet-activating factor in rat liver, *J. Biol. Chem.* **262**, 527–530.

Moolenaar, W. H. (1994) LPA: a novel lipid mediator with diverse biological actions, *Trends Cell Biol.* **4**, 213–219.

Mueller, H. W., O'Flaherty, J. T., and Wykle, R. L. (1984) The molecular species distribution of platelet-activating factor synthesized by rabbit and human neutrophils, *J. Biol. Chem.* **259**, 14554–14559.

Nakayama, R. and Saito, K. (1989) Presence of 1-*O*-alk-1′-enyl-2-*O*-acetyl glycerophosphocholine (vinyl form of PAF) in perfused rat and guinea pig hearts, *J. Biochem.* **105**, 494–496.

Robinson, N. C. (1990) Silicic acid HPLC of cardiolipin, mono- and di-lyso cardiolipin and several of their chemical derivatives, *J. Lipid Res.* **31,** 1513–1516.

Robinson, N. C., Zborowski, J., and Talbert, L. H. (1990) Cardiolipin-depleted bovine heart cytochrome c oxidase: binding stoichiometry and affinity for cardiolipin derivatives, *Biochemistry* **29,** 8962–8969.

Saito, K. (1992) Platelet-activating factor and related compounds, in *Mass Spectrometry: Clinical and Biomedical Applications,* Vol. 1, D. M. Desiderio, editor, Plenum Press, New York, pp. 319–343.

Satouchi, K., Pinckard, R. N., and Hanahan, D. J. (1981) Influence of alkyl ether chain length of acetylglyceryl ether phosphorycholine and its ethanolamine analog on biological activity towards rabbit platelet, *Arch. Biochem. Biophys.* **211,** 683–688.

Satsangi, R. K., Ludwing, J. C., Weintraub, S. T., and Pinckard, R. N. (1989) A novel method for the analysis of platelet-activating factor: direct derivation of glycerophosphalipids, *J. Lipid Res.* **30,** 929–937.

Shimizu, T., Honda, Z., Nakamura, M., Bito, H., and Izumi, T. (1992) Platelet-activating factor receptor and signal transduction, *Biochem. Pharmacol.* **44,** 1001–1008.

Silvestro, L., DaCol, R., Scappaticca, E., Libertucci, D., Biancone, L., and Camussi, G. (1993) Development of a high-performance liquid chromatographic–mass spectrometric technique, with an ionspray interface, for the determination of platelet-activating factor (PAF) and lyso-PAF in biological samples, *J. Chromatogr.* **647,** 261–269.

Smaal, E. B., Romijn, D., van Kessel, W. S. M. Guerts, de Kruijff, B., and de Gier, J. (1985) Isolation and purification of cardiolipin from beef heart, *J. Lipid Res.* **26,** 634–637.

Stoffel, W., Assmann, G., and Binczek, E. (1970) Metabolism of sphingosine bases. XIII. Enzymatic synthesis of 1-phosphate esters of 4t-sphingenine (sphingosine), sphinganine (dihydrosphingosine), 4-hydroxysphinganine (phytosphingosine) and 3-dehydrosphinganine by erythrocytes, *Hoppe-Seyler's Z. Physiol. Chem.* **351,** 635–642.

Sugiura, T., Tokumura, A., Gregory, L., Nouchi, T., Weintraub, S. T., and Hanahan, D. J. (1994) Biochemical characterization of the interaction of lipid phosphoric acids with human platelets: comparison with platelet activating factor, *Arch. Biochem. Biophys.* **311,** 358–368.

Tessner, T. and Wykle, R. (1987) Stimulated neutrophils produce an ethanolamine analog of platelet-activating factor, *J. Biol. Chem.* **262,** 12660–12664.

Tokumura, A., Takauchi, K., Asai, T., Kaniyasu, K., Ogawa, T., and Tsukatani, H. (1989) Novel molecular analogues of phosphatidyl cholines in a lipid extract from bovine brain: 1-long chain acyl-2-short chain acyl-*sn*-glycero-3-phosphocholines, *J. Lipid Res.* **30,** 219–224.

Van Veldhoven, P. P., Foglesong, R. J., and Bell, R. M. (1989) A facile enzymatic synthesis of sphingosine-1-phosphate and dihydrosphingosine-1-phosphate, *J. Lipid Res.* **30,** 611–616.

Venable, M. E., Zimmerman, G. A., McIntyre, T. M., and Prescott, S. M. (1993) Platelet-activating factor: a phospholipid autacoid with diverse actions, *J. Lipid Res.* **34,** 691–702.

Weintraub, S. T., Lear, C. S., and Pinckard, R. N. (1990) Analysis of platelet-activating factor by GC-MS after direct derivatization with pentafluorobenzoyl chloride and heptafluorobutyric anhydride, *J. Lipid Res.* **31,** 719–725.

Weintraub, S. T., Pinckard, R. N., Heath, T. G., and Gage, D. A. (1991) Novel mass spectral fragmentation of heptlafluoro butyryl derivatives of acyl analogues of platelet-activating factor, *J. Am. Soc. Mass Spectrum* **2,** 476–482.

Zhang, H., Desai, N. N., Olivera, A., Seki, T., Brooker, G., and Speigel, S. (1991) Sphingosine-1-phosphate, a novel lipid involved in cellular proliferation, *J. Cell Biol.* **114,** 155–167.

INDEX

208